Khanh Dang Vu

Production de biopesticide bactérien en utilisant des eaux usées

Khanh Dang Vu

Production de biopesticide bactérien en utilisant des eaux usées

Production d'un biopesticide à base de Bacillus thuringiensis en utilisant les eaux usées d'industries d' amidon

Presses Académiques Francophones

Impressum / Mentions légales
Bibliografische Information der Deutschen Nationalbibliothek: Die Deutsche Nationalbibliothek verzeichnet diese Publikation in der Deutschen Nationalbibliografie; detaillierte bibliografische Daten sind im Internet über http://dnb.d-nb.de abrufbar.

Information bibliographique publiée par la Deutsche Nationalbibliothek: La Deutsche Nationalbibliothek inscrit cette publication à la Deutsche Nationalbibliografie; des données bibliographiques détaillées sont disponibles sur internet à l'adresse http://dnb.d-nb.de.

Coverbild / Photo de couverture: www.ingimage.com

Verlag / Editeur:
Presses Académiques Francophones
ist ein Imprint der / est une marque déposée de
OmniScriptum GmbH & Co. KG
Heinrich-Böcking-Str. 6-8, 66121 Saarbrücken, Deutschland / Allemagne
Email: info@presses-academiques.com

Herstellung: siehe letzte Seite /
Impression: voir la dernière page
ISBN: 978-3-8381-4159-6

Zugl. / Agréé par: Québec, Institut national de la recherche scientifique, Diss., 2009

ACKNOWLEDGMENTS

To finish this book as a main part of my Ph.D thesis, I am grateful to many people, and on this occasion, I would like to send my sincere thanks to all of them. First, to my Ph.D. supervisors Prof. R.D. Tyagi and Prof. J.R. Valéro for their exceptional contribution to my knowledge in this domain of research. Their sincere supports, encouragement and untiring supervision throughout my PhD process are very important for my success in finishing my whole researching works.

My thanks are sent to Prof. J.P. Jones, Prof. M. Sirois, Prof. J.F. Blais and Prof. S.K. Brar (INRS-ETE) for their acceptation in judging and evaluating my thesis, their valuable comments to improve the quality of the dissertation are appreciated.

Dr. LeQuoc Sinh, Prof. Nguyen Tien Thang, Prof. Nguyen Thi Quynh, Dr. Hoang Nghia Son, Dr. Vu Van Do, Dr. S. Barnabé, Dr. S.K. Brar, Dr. M. Verma and Dr. S. Yan are acknowledged for their moral support to me throughout my research. Mrs. S. Dussault, Mrs. J. Desrosiers, Mr. S. Prémont, Mrs. M.G. Bordeleau, Mr. R. Rodrigues, Mr. S. Durand, Ms. A. Bensadoune are thanksed for their help and instruction in administrative works and laboratory techniques during my Ph.D. process.

I am grateful to ADM-Ogilvie (Candiac, Québec, Canada) for providing starch industry wastewater; and Insect Production Unit, Great Lakes Forest Research Centre (Sault Ste. Marie, Ontario) for providing spruce budworm larvae during this research.

Dr. Monique Lacroix, Professor at INRS-Institut Armand-Frappier, is acknowledged for her encouragement and advices during the period that I prepared this book.

My sincerece thanks are also sent to my friends/ colleagues (Kokou, Bala, Mathieu, Jean-Philippe, etc.) for their helps, their supports, etc. during the time I conducted the research at INRS. My trainees Aida Ben Aribi, Sabrina Saltaji their supports in preparation of this book for a publication.

Last but not the least, I am greatly grateful to all members of my grand family; especially, Da Van Vu, my father, and Mai Thi Thu Truong, my mother, and my sisters, my brother as well as their own family, for their great loves, moral supports and encouragement in all my life steps. My sincere thanks are also sent to my wife's family and my small family (Doan Huynh Giang, my wife; Vu Doan Quynh Hoa, my daughter; and Vu Gia-Uy, my son) who always displayed their great loves, supports, and encouragement during this memorable journey.

Finally, the book can not be perfect and can have many errors or mistakes. Please give me your advices and opinion for improving it.

Many thanks to all readers,

Québec, 12th, August, 2014

Khanh Dang Vu, Ph.D.

Table of content

List of figures

List of Table

CHAPITRE 1. SYNTHÈSE

1.1. INTRODUCTION

Les insectes ravageurs réduisent considérablement les rendements des cultures agricoles et sylvicoles à travers le monde. Par exemple, au Canada, les larves de la tordeuse des bourgeons de l'épinette ont ravagé des millions d'hectares de forêts de sapins baumiers, d'épinettes blanches, d'épinettes de Norvège, d'épinettes noires, d'épinettes rouges, de pins gris, etc. (Ressources Naturelles Canada). C'est aussi le cas au Vietnam, un pays agricole, où plusieurs types d'insectes nuisibles causent des pertes considérables dans les cultures importantes telles que le riz, le maïs, le thé, le café, le chou, etc. (Nguyen Cong Hao et al., 1996).

Pour contrôler les organismes nuisibles, les pesticides chimiques ont longtemps été employés. Toutefois leur application aveugle et irréfléchie a favorisé la sélection de souches d'insectes nuisibles agricoles ou vecteurs des maladies humaines et provoqué la dégradation de l'environnement. Deux propriétés des produits chimiques, soit l'action résiduelle prolongée et la toxicité envers une gamme étendue d'organismes ont donc provoqué de sérieux problèmes. Ainsi, un besoin urgent de pesticides compatibles avec l'environnement est essentiel pour réduire la contamination et la résistance des insectes contre les pesticides (Ben-Dov et al., 1997). L'utilisation de *Bacillus thuringiensis* (Bt) comme insecticide commercial est justifiée par sa capacité remarquable à produire de grandes quantités de protéines larvicides (connues sous le nom de protéine Cry ou delta-endotoxine) contenues dans les inclusions cristallines parasporales formées durant la sporulation de Bt (Schnepf et al., 1998). La diversité des protéines cristallines insecticides synthétisées par différentes souches de Bt a permis son utilisation pour la lutte biologique en agriculture, en sylviculture et pour la santé publique. Les récentes tendances suggèrent que la lutte biologique est de plus en plus importante dans les stratégies pour une gestion intégrée des organismes nuisibles (Ben-Dov et al., 1997).

Cependant, l'utilisation des biopesticides à base de Bt est limitée en raison de leur prix élevé qui s'explique entre autres par leur coût élevé de production. À ce propos, le coût de la matière première pour la production de Bt représente jusqu'à 40% du coût global de production (Lisanky et al., 1993). Par conséquent, l'emploi de matières premières alternatives et économiques est très important. Dans cette perspective, le projet intitulé Bt-INRS dans lequel les boues d'épuration des eaux usées municipales sont employées comme matières premières alternatives de fermentation a montré des résultats forts prometteurs en termes d'entomotoxicité, soit le principal paramètre pour évaluer l'efficacité des biopesticides à base de Bt. Ce projet a spécialement démontré que l'emploi des boues d'épuration comme matière première pour la production de biopesticides à base de *B. thuringiensis* var. *kurstaki* HD-1 (Btk HD-1) présente de nombreux avantages en termes d'économie, d'efficacité et de formulation (Brar, 2007). Par exemple, en employant les boues d'épuration (non pré-traitées) comme matière première pour la production de Btk HD-1, l'entomotoxicité obtenue était généralement plus élevée que celle obtenue avec le milieu de culture conventionnel. L'activité insecticide obtenue est aussi supérieure lors de l'application de certaines techniques telles que la fermentation en fed-batch (Yezza et al., 2005), l'addition de Tween 80 dans les boues d'épuration brutes (Brar et al., 2005), le pré-traitement des boues pour augmenter leur biodégradabilité (Barnabe, 2005 ; Yezza et al., 2006), l'addition de différentes sources de carbone et/ou d'azote dans les boues d'épuration (LeBlanc, 2004), etc.

Les eaux usées d'industrie d'amidon (**SIW**) sont un des résidus provenant des eaux usées industrielles produites en grandes quantités dans le monde (Rajbhandari et Annachhatre, 2004). SIW contient une forte concentration de la demande chimique en oxygène (COD), sous la forme de carbone et de sources d'azote (amidon, gluten, protéines, fibres et autres minéraux), ce qui pose un sérieux problème de traitement et de leur élimination finale dans l'environnement. Dans ce but la pratique générale est le traitement soit en aérobie (Malladi et Ingham, 1993; Rajasimman et Karthikeyan, 2007) soit en anaérobie (Annachhatre et Amatya, 2000; Rajbhandari et Annachhatre, 2004, Colin et al., 2007) de SIW avec plus ou moins d'efficacité . Une autre approche pour résoudre les problèmes environnementaux des SIW est le traitement par la biotechnologie verte. Dans cette approche, les éléments nutritifs présents dans SIW sont recyclés par voie biotechnologique en produits à haute valeur ajoutée telles qu'une biomasse microbienne et enzymatique (Jin et al., 1999, 2002), des agents de lutte biologique (Verma et al., 2006) et des biopesticides (Brar et al., 2005, Yezza et al., 2006).

Dans des recherches sur l'utilisation des SIW comme matières premières pour la production de biopesticides à base de Btk HD-1 (Brar et al., 2005, Yezza et al., 2006), les résultats démontrent les avantages par rapport à l'utilisation des boues: (1) Les SIW peuvent être employées directement sans pré-traitement et l'entomotoxicité obtenue est souvent plus élevée que celle constatée avec les boues d'épuration (Yezza et al., 2006); (2) le bouillon fermenté présente plusieurs avantages lors de l'étape de la formulation (Brar et al., 2005). Cependant, la recherche systématique sur la maximisation de l'entomotoxicité de biopesticides à base de Btk HD-1 en utilisant comme matières premières n'est pas encore totalement explorée.

1.2. REVUE DE LITTÉRATURE

1.2.1. Bacillus thuringiensis et les mécanismes d'action des biopesticides-Bt contre les insectes nuisibles

1.2.1.1. Bacillus thuringiensis: classification et distribution

1.2.1.1.1. Qu'est-ce que Bacillus thuringiensis (Bt) ?

Bt est une bactérie en forme de bâtonnet, aérobie, gram-positif, sporulée qui forme un cristal protéique parasporal au cours de la phase stationnaire de son cycle de croissance (Figure 1). Bt a d'abord été caractérisé comme un agent entomopathogène et son activité insecticide a été attribuée en grande partie ou entièrement (en fonction de l'insecte) aux cristaux parasporaux (Schnepf et al., 1998). Ces inclusions parasporales contenant les cristaux de protéiques insecticides ou δ-endotoxines (Cry toxines: Nomenclature de toxines de Bt), qui sont codées par des gènes *cry*. Cry toxines ont des activités insecticides contre les espèces de l'ordre des Lépidoptères, et contre certains larves de diptères (mouches et moustiques) et de coléoptères (scarabées) (Schnepf et al., 1998). Ainsi, Bt synthétise un important réservoir de toxines Cry et des gènes *cry* pour la production d'insecticides biologiques et éventuellement de cultures génétiquement modifiées et résistantes aux insectes nuisibles. Une brève histoire de la recherche sur Bt est présentée dans le tableau 1 (Krieg et al., 1983; Schnepf et Whiteley, 1981; Estruch et al, 1996; Mizuki et al., 2000; Federici, 2005;).

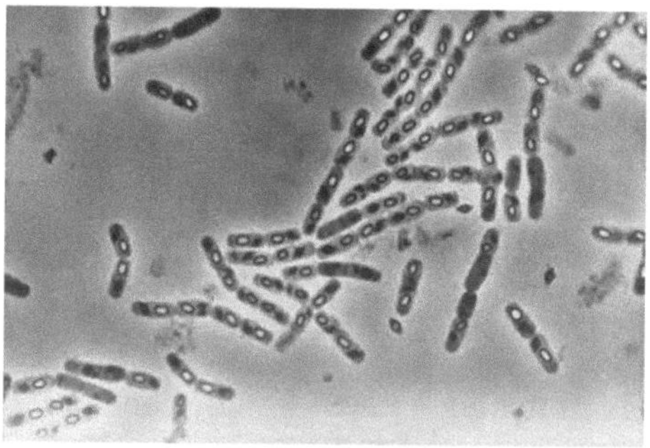

Figure 1. Colonie de *Bacillus thuringiensis*

Source :http://www.futurasciences.com/uploads/tx_oxcsfutura/comprendre/d/images/604/pintureau_03.jpg

Tableau 1. Une brève histoire de la recherche sur le Bt

Année	Chercheur(s)/Compagnie	Événement
1901	Ishiwata	Découverte de cette bactérie connue plus tard sous le nom de *Bacillus thuringiensis*, au Japon, au cours de son étude d'une maladie bactérienne de vers à soie.
1915	Berliner	Découvre un bacille similaire qui tue les larves de la noctuelle de farine, en Allemagne. Il a publié une description de la bactérie et de ses propriétés, et l'a nommée *Bacillus thuringiensis*. Le nom de l'espèce "*thuringiensis*» est dérivé de la Thuringe, région en Allemagne, dans laquelle la maladie a eu observée.
1930s	En France	Développement d'un produit à base de Bt pour utilisation sur les cultures maraîchères pour contrôler les chenilles nuisibles et finalement développé un produit appelé "Sporeine" à la fin des années 1930.
1951	Steinhaus	Résurrection de l'intérêt concernant Bt pour son utilisation comme insecticide commercial.
1953	Hannay	Études relatives aux corps cristallins parasporaux de Bt
1954	Angus	Démontre que les cristaux de Bt peuvent tuer des vers à soie
1962	De Barjac and Bonnefois	Développement d'un système de stéréotypage des souches de Bt permettant de les distinguer en termes correspondant : subspécifique et sérovariétè type / numéro

1960s	Abbott Laboratories and Sandoz Corporation	Production des biopesticides commerciaux (Dipel et Thuricide) à base de Bt var. *kurstaki* HD-1
1977	Goldberg and Margalit	Bt subsp. *israelensis* (H 14) a été découvert dans le Désert du Néguev en Israël, Bti possède une excellente activité contre les larves d'un grand nombre de moustiques et d'espèces de mouches noires.
1981	Dulmage et Burges	Organisent un groupe international de chercheurs pour utilisation de bioessais normalisés et d'un standard de préparation sur la base des espèces HD 1 de Bt subsp. *kustaki* (H 3a3b) pour comparer de nouvelles souches de Bt
1981	Schnepf and Whiteley	Décodent le premier gène codant pour une protéine insecticide (la protéine Cry) de Bt var. *kurstaki* HD-1 Dipel .
1983	Krieg et al.	Découvrent Bt subsp. *morrisoni* (H 8a8b, variété *tenebrionis*) en Allemagne (communément connu sous le nom de Btt) et qui est active contre les larves et même les adultes de certaines espèces de coléoptères.
1996	Estruch et al.	Découverte de la Vip3A, une nouvelle protéine insecticide végétative de Bt ayant un large spectre d'activités de lutte contre les lépidoptères.
2000	Mizuki *et al*.	Parasporine, une nouvelle protéine avec une activité cytotoxique unique découverte pour la 1ère fois à partir de Bt.

1.2.1.1.2. Classification de Bt

La différenciation des souches de Bt par agglutination flagellaire a été largement utilisée. Depuis son introduction en 1963, le sérotypage a permis de créer un certain ordre dans le groupe Bt , se fondant sur une classification spécifique, stable et fiable de caractère en employant l'antigène H (de Barjac et Frachon, 1990). Cette méthode (dite H-sérotype) regroupe différentes souches de Bt sérovars et présentait beaucoup d'avantages par rapport aux autres procédés de classification de Bt comme le type de bactériophage (Ackermann et al., 1995), la caractérisation biochimique, motif estérase, etc. (de Barjac and Frachon, 1990). En 1999, Lecadet et al. ont révisé et mis à jour la classification des souches de Bt basée sur le sérotype H et 82 sérotypes ont été identifiés jusqu'au sérotype 69 et 13 dans un sous-groupe antigénique (Lecadet et al., 1999). Cette technique de sérotypage, tout en ayant fourni une précieuse base pour le classement de Bt pendant 40 ans, ne fournit aucune information sur la parenté génétique des souches au sein des groupes et entre les groupes. Ainsi, certaines des méthodes de typage des gènes ont été appliquées dans la construction de la relation phylogénétique entre les sérovariétés de Bt.

Par exemple, basé sur les ARN ribosomiques (polymorphismes de longueur des fragments de restriction de gène de l'ARNr 16S [RFLP]) des sérovars de Bt (80 sérovars), la relation phylogénétique a été analysée. Elle montre quatre groupes distincts et deux sérotypes dissocieés ayant 92,5% de parenté en ADN (Joung et Côté, 2001a). En particulier, la combinaison des motifs de restriction de gène de l'ARNr 23S et l'ARNr 5S avec des motifs de restriction du gène de l'ARN 16S a engendré un nouvel arbre phylogénétique des sérovars de Bt (80 sérovars), qui se compose de VIII groupes (Joung et Côté, 2001b). Ainsi, le ribotypage basé sur les RFLP des souches de Bt a permis la différenciation des sérovars, des souches intra-serovars et des souches ne pouvant pas être sérotypées (Joung et Côté, 2001b).

Récemment, l'analyse génotypique des souches de Bt sur la base des ADN polymorphes amplifiés aléatoirement (RAPD) par polymérisation en chaînes (PCR) [RAPD-PCR] a été appliquée pour 126 souches de Bt (56 sérovars), 58 types génomiques ont été relevés (Gaviria Rivera and Priest, 2003). Les motifs de RAPD-PCR obtenus ont révélé que l'espèce Bt est hétérogène. Toutefois, les souches à l'intérieur de certains sérovars sont génomiquement homogènes, tandis que d'autres sérovars de divers groupes, à l'intérieur de ces types de la RAPD, n'ont pas de corrélation avec les autres (Gaviria Rivera and Priest, 2003).

1.2.1.1.3. Distribution de Bt

Bt est une bactérie ubiquitaire largement répandue dans le sol (Bernhard et al. 1997; Vilas-Boas et Franco Lemos, 2004), dans le phylloplane de différentes plantes ou des feuilles (Hansen et al. 1998; Smith et Couche 1991), chez des insectes (Hansen et al., 1998), dans les grains entreposés (Meadows et al. 1992). Au cours des dernières années, Bt a également été retrouvé dans d'autres milieux tels que eaux douces (Ichimatsu et al., 2000) et boues activées (Mizuki et al., 2001, Mohammedi et al., 2006). Lee et al. (2002) ont signalé que Bt est communément associé aux excréments d'animaux, par exemple les herbivores maintenus dans un zoo au Japon, etc.

Les souches obtenues par de nombreux programmes d'isolement sont identifiées par leur gènes *cry* à l'aide la plupart du temps, de méthodes basées sur la PCR utilisant des amorces spécifiques de différents gènes (Ben-Dov et al. 1997; Bravo et al. 1998; Carozzi et al. 1991). Cette voie rapide d'identification des gènes *cry* a augmenté le nombre de gènes, et plus de 250 séquences de gènes *cry* ont été déterminées (http://www.biols.susx.ac.uk/home/Neil_Crickmore/Bt/).

Récemment, dans l'évaluation des effets biologiques des protéines envers les insectes, certaines inclusions parasporales de souches de Bt non-insecticides ont été trouvées, ayant une activité contre des cellules cancéreuses humaines

(Mizuki et al., 1999, 2000, Lee et al. 2000; Ito et al. 2004; Yasutake et al., 2005, Jung et al., 2007). Ces protéines *cry* ont été désignées comme 'parasporines' avec la définition: 'Ce sont des protéines parasporales produites par Bt et d'autres bactéries qui sont non-hémolytique, mais capables de tuer des cellules cancéreuses' (http://parasporin.fitc.pref.fukuoka.jp/intro.html).

1.2.1.2. Formation, structure et rôle des spores de Bt pour son activité insecticide.

1.2.1.2.1. Formation des spores (sporulation)

Bt se cultive facilement sur milieu artificiel. Par exemple, un milieu synthétique défini a été développé pour la culture d'une grande variété de souches de Bt (Nickerson et Bulla, 1974). En milieu liquide, lorsque la population cellulaire a atteint un maximum, les cellules végétatives sporulent. Du point de vue biochimique, la sporulation extensive peut être initiée par la privation de nutriments essentiels, tels que le glucose, l'ammoniaque, phosphate ou autre (Tzeng et Liu, 2000). La formation d'une endospore (l'état de dormance de Bt) est un processus de division cellulaire qui inclut la formation de protoplastes dans le cytoplasme de la cellule (Bulla et al., 1975).

La séquence du développement des spores et de la formation des cristaux parasporaux chez Bt est résumée selon les étapes classiques de la sporulation: la phase I (7 h), la formation de filament axial dans lequel il n'y a pas d'implication apparente des mésosomes avec le nucléole; phase II (7 à 8 h), formation de septum d'avant-spore impliquant les mésosomes; phase III (8 à 9 h), la formation d'engouffrement avec la participation des mésosomes, première apparition d'inclusions ovoïdes et de cristaux parasporaux, changement dans la capacité de la teinte des membranes et du cytoplasme, et formation de la pré-spore; phases IV à VI (9 à 12 h), formation de l'exosporium, paroi cellulaire primordiale, cortex, et manteaux de spore accompagnée par la transformation de la nucléoide de spore; phase VII (après 12 h), maturation des spores (Bechtel et Bulla, 1976). Notez que les spores à maturité ou les spores après la lyse cellulaire (phase VII) ont plusieurs couches distinctes. À partir de la membrane intérieure, elles sont situées comme suit: paroi cellulaire primordiale, cortex, membrane externe, cytoplasme de la cellule mère incorporée, couche de spore lamellaire, couche externe fibreuse de spore, et l'exosporium.

1.2.1.2.2. Rôle de spores et de leur paroi sur les insectes nuisibles

La toxicité des spores de Bt contre *Manduca sexta*, a été démontrée par Schesser et Bulla en 1978. Le pourcentage de mortalité larvaire était corrélé avec le contenu en poids sec des spores. Sur une superficie de 6.8ng/cm^2, il y avait 85% de survie, mais moins de 50% ont survécu à 68.2ng/cm^2. Il se trouve qu'il y a une similitude frappante entre l'action des spores et celle des cristaux parasporaux, c'est-à-dire l'inhibition de la croissance du sphinx du tabac. Les effets entomotoxiques des spores de Bt contre les larves d'insectes sont reliés à l'action de leur paroi qui contient la même composante que la pro-toxine cristalline des cristaux parasporaux (Schesser et Bulla, 1978, Johnson et al., 1998).

Dans certains cas, les spores de Bt sont impliquées dans son entomopathogénicité (Dubois & Dean, 1995, Johnson & McGaughey, 1996, Li et al., 1987; Liu et al., 1998). Il a été suggéré que l'effet toxémique des cristaux protéiques crée des conditions favorables au développement de la bactérie dans l'intestin des larves d'insectes affaiblies (Schnepf et al., 1998). Ensuite, les bactéries peuvent envahir l'hémolymphe et provoquer une septicémie. Il est probable que les spores agissent en synergie avec la toxine et celle-ci peut aussi résulter de leur germination dans l'intestin moyen avec la production d'une variété de facteurs pathogènes par les cellules végétatives tels que phospholipase C, hémolysines et entérotoxines (Salamitou et al.,

2000), y compris certaines Protéines Insecticides Végétatives, les VIP (Estruch et al., 1996; Aronson, 2002).

1.2.1.3. Synthèse des delta-endotoxines (protéines Cry) actives contre les lépidoptères

1.2.1.3.1. Gènes codant pour des protéines Cry

Les gènes codant des protéines Cry ont été nommés gènes *cry* et leurs emplacements de synthèse situés sur des plasmides de Bt (Whiteley et Schnepf, 1986; Aronson et al., 1986). En 1989, Höfte et Whiteley ont examiné les gènes *cry* connus et en ont proposé une nomenclature systématique (Höfte et Whiteley, 1989). Depuis, le nombre de séquences des gènes (codant pour les protéines Cry et Cyt) est passé de 14 à plus de 250 (Schnepf et al., 1998; Crickmore et al., 1998; http://www.biols.susx.ac.uk/home/Neil_Crickmore/Bt/). Toutefois, les gènes *cry* les plus connus (codant pour des protéines Cry) sont *cry1* (action spécifique contre les lépidoptères), *cry2* (action spécifique contre les lépidoptères et diptères), *cry3* (action spécifique contre les Coléoptères), *cry4* (action spécifique contre Diptères) (Höfte et Whiteley, 1989; Schnepf et al., 1998; Crickmore et al., 1998). De nombreuses recherches ont aussi porté sur le clonage des gènes *cry1*, sur les mécanismes d'action et sur l'expression de ces gènes chez les plantes (Höfte et Whiteley, 1989; Schnepf et al., 1998; Crickmore et al., 1998).

Certaines souches de Bt ne produisent qu'un seul type de protéines Cry. Par exemple, Bt var. *kurstaki* HD-73 ne produit que Cry1Ac. De nombreuses souches de Bt produisent plus d'un type de protéines Cry; Chacune ayant sa propre spécificité insecticide comme par exemple, la souche HD-1 de Bt var. *kurstaki* qui contient des gènes codant pour différentes protéines Cry: Cry1Aa, Cry1Ab, Cry1Ac (spécifique à Lépidoptères) et Cry2Aa (spécifiques aux lépidoptères et diptères) et Cry2Ab (spécifique à Lépidoptères) (Widner et Whiteley, 1989; Höfte et Whiteley, 1989).

En fait, il est difficile de prédire l'activité biopesticide des souches de Bt en se basant seulement sur les gènes *cry* ou les produits de PCR de souches de Bt (Porcar et Juárez-Pérez, 2003). Par exemple, bien que les protéines de type Cry1A ont 90% de séquences identiques en acides aminés, différentes spécificités d'insecticide ont été signalées contre divers lépidoptères (tableau 2) (Höfte et Whiteley, 1989).

Les résultats dans le tableau 2 montrent que chaque type de protéine Cry1A a une valeur létale différente en termes de CL_{50} sur les espèces d'insectes : La valeur dépendant de l'espèce. Pour une même espèce d'insectes, les valeurs de CL_{50} sont également différentes. Trois types de protéines Cry1A (Cry1Aa, Cry1Ab et Cry1Ac) sont structurellement très proches, cependant, elles montrent un large spectre d'activités bioinsecticides qui se chevauchent. En résumé, seuls les essais biologiques avec des cristaux purifiés sont en corrélation avec la teneur en protéines Cry de chaque souche, et en partie avec le contenu de gènes cry et cyt (Porcar et Juárez-Pérez, 2003).

1.2.1.3.2. Synthèse des protéines Cry et structure des cristaux protéiques parasporaux de Bt

Ces différentes protéines insecticides (protéines Cry) sont synthétisées au cours de la phase stationnaire de croissance et s'accumulent dans la cellule mère dans des inclusions cristallines parasporales, ce qui peut représenter jusqu'à 25% du poids sec des cellules sporulées (Figure 2). La quantité de protéines cristallines produite par une culture de Bt en conditions de laboratoire (environ 0,5 mg de protéines par ml) et la taille des cristaux indiquent que chaque cellule doit synthétiser environ 10^6 à $2x10^6$ de molécules de delta-endotoxines lors de la phase stationnaire et ce sous la forme de cristaux (Agaisse et Lereclus, 1995).

Tableau 2. Toxicité des protéines cristallines contre les espèces de lépidoptères

protéines Cry	Souches et sous-souches de Bt	CL_{50} de protéines pour:				
		Pieris brassicae (µg/ml)	*Mandaca sexta* (ng/cm^2)	*Heliothis virescens* (ng/cm^2)	*Mamestra brassicae* (ng/cm^2)	*Spodoptera littoralis* (ng/cm^2)
Cry1Aa	*azawai* HD-68	0.8	5.2	90	165	>1350
Cry1Ab	*berliner* 1715	0.7	8.6	10	162	>1350
Cry1Ac	*kurstaki* HD-73	0.3	5.3	1.6	2000	>1350

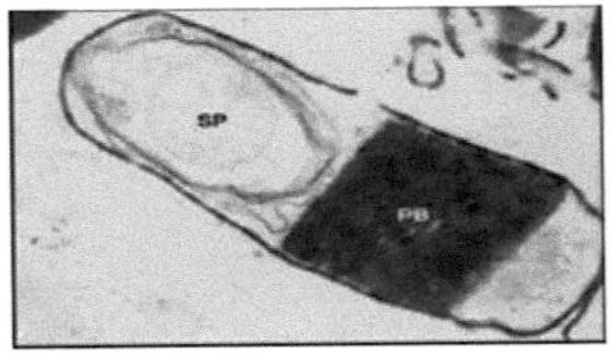

Figure 2. Microscopie électronique de *Bacillus thurigniensis* au cours de la sporulation. L'inclusion parasporale noire est le cristal insecticide

Source: http://www.ufrgs.br/laprotox/digestion-eng.htm

Les inclusions parasporales sont développées au cours de phases III à VI (8 à 12 h) de la sporulation. Le cristal parasporal de Bt a été observé durant l'engouffrement (phase III, 8 h) et possède un aspect cristallin à cette étape du développement. Le cristal atteint presque sa pleine grandeur au moment où l'exosporium apparaît (étape IV, 9 h). Un ovoïde pourrait être développé en même temps que l'apparence du cristal, Toutefois, certaines cellules ne produisent pas cette inclusion. L'inclusion ovoïde est facilement distinguable du cristal parasporal (forme de diamant) parce qu'elle n'est pas cristalline et apparaît toujours ovoïde.

Plusieurs cristaux peuvent être synthétisés au sein d'une cellule, mais une seule inclusion ovoïde par sporange a été observée. Par ailleurs le cristal parasporal et l'inclusion ovoïde ne sont pas toujours associés à une structure spécifique. Des sections sérielles révèlent que le cristal n'est pas connecté aux mésosomes ou à des membranes provenant de la forespore, ou de l'exosporium (Bechtel et Bulla, 1976).

1.2.1.3.3. Mode d'action de la toxine Cry1A contre les Lépidotères

En général, les toxines Cry qui constituent les inclusions cristallines sont suffisamment létales pour tuer les larves d'insectes infectés. Lorsqu'ils sont ingérés par les larves d'insectes sensibles, ces cristaux de protéines sont dissous et activés dans le tube digestif des insectes. Ils se lient à des récepteurs spécifiques situés sur les cellules épithéliales de l'intestin moyen, provoquant des pores transmembranaires et causant la lyse cellulaire. Cette toxémie peut entraîner, par elle-même la mort des larves d'insectes (Grochulski et al., 1995; Agaisse et al., 1999). Le mécanisme d'action des protéines Cry de Bt est présenté dans la Figure 3 (modifié à partir de Whalon et Wingerd, 2003).

En particulier, les toxines Cry1A produites comme inclusions cristallines doivent être ingérées par les larves sensibles pour y être toxiques. Ces cristaux sont dissous dans le milieu alcalin et réductif de l'intestin moyen des larves, ce qui libère des pro-toxines solubles de 130 kDa et les pro-toxines inactives solubilisées sont clivées par des protéases de l'intestin moyen en fragments de 60-70 kDa contenant des protéines résistantes aux protéases (Bravo et al., 2007).

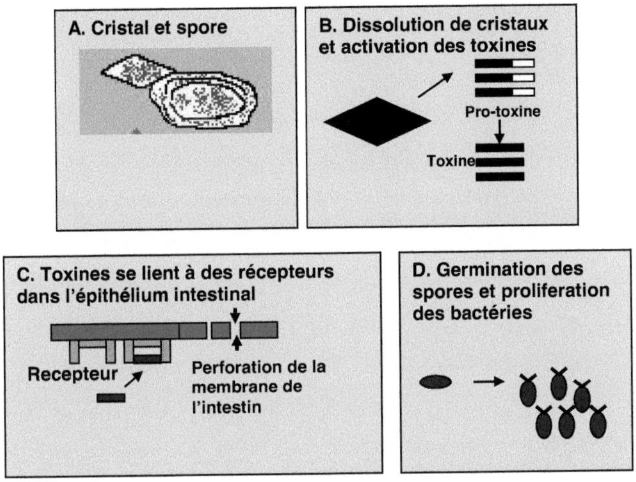

Figure 3.Mécanisme de la toxicité des protéines Cry

A: L'ingestion de spores ou de protéines recombinantes par des larves phytophages. B: Dans l'intestin moyen, les endotoxines sont solubilisées à partir des inclusions de protéines cristallisées. Pro-toxine Cry sont traitées protéolytiquement en toxines actives dans l'intestin moyen. C: Toxine active se lie à des récepteurs sur la surface des cellules épithéliales cylindriques pour former des pores dans la membrane. D: lourds dommages à la membrane de l'intestin moyen conduisant à la famine ou à la septicémie.

L'activation de la toxine implique l'élimination protéolytique d'un peptide N terminal (25-30 acides aminés pour Cry1 toxines, 58 résidus pour Cry3A et 49 résidus pour Cry2Aa) et environ la moitié des protéines reste à partie du C-terminal dans le cas des longues pro-toxines Cry. La toxine activée se lie à des récepteurs spécifiques sur la membrane de la bordure en brosse des cellules cylindriques de l'épithélium de l'intestin moyen (de Maagd et al., 2001; Bravo et al., 2007) avant de s'insérer dans la membrane. L'insertion de la toxine entraîne la formation de pores lytiques dans les microvillosités des membranes apicales (Aronson et Shai, 2001; Bravo et al., 2007). Ultérieurement, la lyse cellulaire et la perturbation de l'épithélium de l'intestin moyen libèrent les composantes cellulaires permettant la germination des spores qui provoque une septicémie sévère et la mort des insectes (de Maagd et al., 2001; Bravo et al., 2007).

1.2.1.3.4. Structure et fonction des toxines Cry1Aa spécifiques aux lépidoptères ravageurs

Cry1Aa possède trois domaines. Le domaine I consiste en un ensemble de sept α-hélices antiparallèles dans laquelle l'hélice 5 est entourée par le reste des hélices. Le domaine II se compose de trois feuillets β antiparallèles joints dans une topologie typique "Greek clé", disposés dans un pli en β-prisme (330, 343). Le domaine III se compose de deux feuillets β tordus, antiparallèles formant un sandwich β avec un topologie "Jelly Roll" (Grochulski et al., 1995; Schnepf et al., 1998). La présence des hélices amphipathiques et hydrophobes du domaine I, indique que ce domaine pourrait être responsable de la formation des pores lytiques dans l'épithélium intestinal de l'organisme cible, un des mécanismes proposés pour l'activité toxique de Cry (Grochulski et al., 1995; Schnepf et al., 1998). Les boucles exposées à la surface d l'apex des trois feuillets β du domaine II ont d'abord été présentées comme candidats pour la capacité à lier les récepteurs (Grochulski et al., 1995; Schnepf et al., 1998). La structure β-sandwich du domaine III pourrait jouer un certain nombre de rôles clés dans la biochimie de la toxine. Le domaine III a été proposé pour la conservation des fonctions pour le maintien de l'intégrité structurale de la toxine, peut-être en la protégeant contre la protéolyse dans le tube digestif de l'organisme cible (Li et al., 1991; Schnepf et al., 1998). Il a été suggéré qu'il joue le rôle de lien à des récepteurs dans certains systèmes (Schnepf et al., 1998).

1.2.1.3.5. Liaison des toxines à des récepteurs de larves de lépidoptères

Pour les toxines Cry1A , au moins quatre protéines de liaison ont été décrites dans différents insectes lépidoptères: un glycosylphosphatidyl-inositol (GPI)-ancrée-aminopeptidase N (APN) 120 kDa chez *Manduca sexta* (Knight et al., 1994); Une protéine ressemblant à la cadhérine (CADR) chez *Manduca sexta* (Vadlamudi et al., 1995); une phosphatase alcaline (ALP) ancre- GPI chez *Heliothis virescens* (Jurat-Adang et Fuentes, 2004); et un glycoconjugué (270 kDa) chez la spongieuse (Valaitis et al., 2001).

La figure 4 représente les quatre types de récepteurs putatifs des molécules Cry1A caractérisées à ce jour (modifié à partir de Bravo et al., 2007). Ainsi, la liaison de la protéine Cry1A avec des protéines de liaison (ou récepteurs) cause l'insertion de la toxine dans la membrane de bordure en brosse des cellules cylindriques de l'épithélium de l'intestin moyen, ce qui conduit à la formation de pores lytiques dans les membranes apicales des microvillosités suivie de la lyse cellulaire conduisant à la mort de l'insecte. Les détails d'information ont été présentés dans de nombreuses revues (Schnepf et al., 1998; Bravo et al., 2007; Gómez et al., 2007).

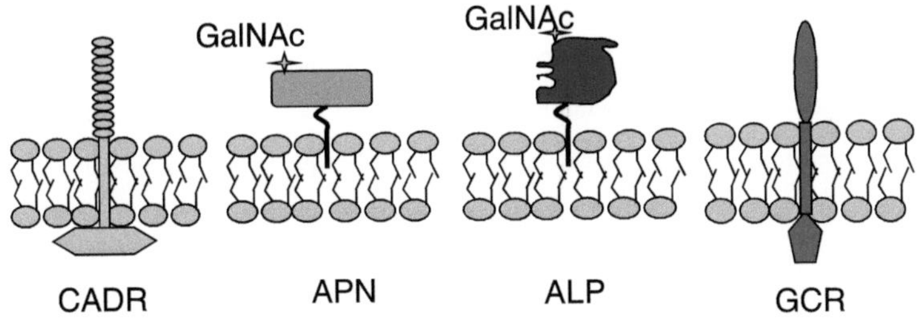

Figure 4.Molécules réceptrices de protéines Cry1a

CADR, cadhérine récepteur; APN, aminopeptidase-N; ALP, phosphatase alcaline; GCR, récepteur 270 kDa glyco-conjugué

1.2.1.4. Synthèse d'autres éléments par Bt au cours de sa croissance et rôle de ces éléments dans le pouvoir entomotoxique du biopesticide

En plus des protéines Cry, Bt produit différents facteurs de virulence, y compris les protéines insecticides sécrétées, des α-exotoxines (phospholipase C ou lécithinase), des β-exotoxines, des hémolysines, des entérotoxines, des chitinases, etc. Tous ces facteurs peuvent avoir un rôle dans le pouvoir entomopathogène contre les insectes dans des conditions naturelles, et contribuer au développement de la bactérie dans les larves d'insectes mortes. Toutefois la contribution exacte de chaque facteur est souvent inconnue (Schnepf et al., 1998; Agaisse et al., 1999; de Maagd et al., 2001; Gohar et al., 2005). Ces facteurs peuvent être de grandes molécules de hauts poids moléculaires (protéines) ou de petites molécules. Certaines n'étant pas acceptables dans les préparations commerciales de biopesticides à base de Bt (Tableau 3).

1.2.1.4.1. Protéines insecticides végétatives (VIPs)

Il a été observé que certaines souches de Bt produisent des protéines insecticides végétatives (VIPS). Ces VIPS présentes dans le surnageant de culture pendant la croissance végétative de certains Bt, sont des protéines insecticides d'un genre nouveau qui ne présentent aucune homologie avec des cristaux protéines insecticides (ICPs) (Estruch et al. 1996). Ces protéines comprennent les toxines binaires Vip1/Vip2 , qui sont toxiques pour les coléoptères, et Vip3 qui est toxique pour les lépidoptères (Estruch et al. 1996; Warren 1998).

Il a été démontré que Vip3A est toxique contre un large spectre d'insectes lépidoptères dont *Agrotis ipsilon, Spodoptera exigua, Chilo partellus, Helicoverpa punctigera and Diatraea saccharalis* (Estruch et al. 1996; Donovan et al. 2001; Selvapandiyan et al. 2001; Loguercio et al. 2002). Des différences ont été trouvées dans le mode d'action de Vip3A par rapport à l'effet des protéines Cry. L'apparition de symptômes provoqués par l'ingestion de Vip3A survient 36-48 h plus tard par rapport à celui de la delta-endotoxines (Yu et al. 1997). En outre, le Vip3A se lie sur des récepteurs différents sur la bordure en brosse des vésicules membranaires (BBMV), en comparaison avec Cry1A protéines (Lee et al. 2003). Les VIPS peuvent être utilisés séparément comme biopesticides (Estruch et al. 1996).

1.2.1.4.2. Chitinases

Depuis longtemps, les chitinases, enzymes hydrolysant la chitine (un homopolymère de β-1 ,4-N-acétyl-alignés glucosamine), ont été utilisées comme agent de synergie pour accroître l'entomotoxicité de biopesticides (Smirnoff and Valéro, 1972; Smirnoff, 1973, 1974). Il a été suggéré que la chitinase pourrait accroître la toxicité du Bt par perforation de la barrière de la membrane péritrophique de l'intestin moyen des larves et donc d'accroître l'accessibilité de la delta-endotoxines de Bt aux récepteurs sur les membranes des cellules épithéliales (Regev et al. 1996). La perforation de la membrane péritrophique par la chitinase a été clairement démontrée par l'examen "in vivo" et "in vitro" par les études de Wiwat et al., (2000); Thamthiankul et al., (2004).

Ces résultats confirment le rôle essentiel de la chitinase dans l'hydrolyse de la membrane peritrophique, permettant ainsi la pénétration des spores et des cristaux de Bt dans l'hémolymphe des larves. Au cours des dernières années, il y a eu quelques recherches sur : (1) La sélection de souches de Bt qui peuvent produire des chitinase (Rojas-Avelizapa et al., 1999; Barboza-Corona et al., 1999; Liu et al., 2002); (2) Le clonage et l'expression de gènes codant pour la chitinase (provenant d'autres microbes) en Bt pour l'accroissement de l'entomotoxicité (Wiwat et al., 1996; Tantimavanich et al., 1997; Lertcanawanichakul et Wiwat, 2000; Sirichotpakorn et al, 2001; Tantimavanich et al., 2004; Thamthiankul et al., 2004); (3) La production des chitinases endogènes ainsi que le clonage, le séquençage des gènes codant pour la chitinase ou les caractéristiques de certains souches de Bt, tels que Bt var. *kurstaki* (Wiwat et al., 2000; Arora et al., 2003; Driss et al, 2005), Bt var. *parkitani* (Thamthiankul et al., 2001), Bt var. *kenyae* (Barboza-Corona et al, 2003), Bt var. *israelensis* (Zhong et al, 2003), Bt var. *alesti* (Lin and Xiong, 2004), Bt var. *sotto* (Zhong et al., 2005).

Toutefois, ces recherches ont été le plus souvent utilisées en fermentation en milieu synthétique complétée par de la chitine colloïdale pour la production de chitinase et de biopesticides à base de Bt sauf pour les recherches de Rojas-Avelizapa et al., 1999, qui ont été appliqués pour fermentation à l'aide de déchets (déchets blanchis de crevettes) comme seul ingrédient pour la production de chitinases et de biopesticides de Bt.

1.2.1.4.3. Protéases de Bt

Pendant la croissance, Bt peut produire certains types de protéase nécessaires pour hydrolyser des protéines dans les milieux (synthétiques ou complexes). Ainsi, une métalloprotéase (37,5 kDa; pH optimal de 6.5-7.5) a été produite par Bt var. *kurstaki* HD-1 (Btk HD-1) dans le milieu synthétiques (Li et Yousten, 1975); quatre protéases ont été fractionnées et caractérisées dans le milieu du bouillon de fermentation (Zouari et Jaoua, 1999); une protéase sérine thermostable (34 kDa, ayant une activité maximale à 70 °C et à un pH de 8.5-9.0) produites par Bt var. *kurstaki* HD-255 en milieu synthétique (Kunitate et al., 1989); deux des protéases neutres et alcalines ont été produites au cours de la croissance de Btk HD-1 par croissance dans les boues d'épuration (Tyagi et al., 2002), etc. En outre, il a été constaté que certains protéases alcalines sont associées à des cristaux protéines insecticides (Bulla et al., 1977; Chestukhina et al., 1980; Nickerson et al., 1981; Bulla et al., 1981, Andrews et al. 1985).

Tableau 3. Exemples de facteurs possibles qui sont produits par Bt au cours de la fermentation

Composants	Caractéristiques moléculaires	Rôle dans les biopesticides	Références
Protéines insecticides végétatives (VIPS)	VIP3A ~ 89 kDa, spécifique à Lépidoptère; VIP1/Vip2 ~ 100/52 kDa, spécifique à Coléoptère	VIPS peuvent être utilisés séparément en tant que biopesticides	Estruch et al., 1996; Warren et al., 1998
Chitinases	Selon les souches de Bt	Les chitinases peuvent améliorer l'entomotoxicité des delta-endotoxines de Bt contre les insectes ravageurs, en agissant sur lamembrane péritrophique	Regev et al., 1996; Sampson and Gooday, 1998; Wiwat et al., 2000; Barboza-Corona et al., 2003; Driss et al., 2005
Protéases alcalines	AprA : protéines ayant une masse moléculaire déduite (précurseurs) de 42.3 kDa.	Peu défini (elles peuvent dégrader les protoxines de Cry mais n'ont pas d'effet	Tan and Donovan, 2000; Brar et al., 2007

Protéases neutre	Npr ayant une masse moléculaire déduite (précurseurs) de 61.0 kDa	direct sur l'activité insecticide)	Donovan et al., 1997; Brar et al., 2007
Inhibiteurs immunitaire A	InA (~ 78 kDa)	Il est actif contre le système de défense humorale des insectes ravageurs	Siden et al., 1979; Dalhammar and Steiner, 1984
Zwittermicin A	396 Da	Quand il est seul, il n'est pas toxique pour les insectes ravageurs, mais il peut être un agent de synergie pour les biopesticides Bt.	Broderick et al., 2000; 2003
Phospholipase C (Lecithinase or alpha-exotoxin)	Une phospholipase C spécifique à phosphatidylinositol a été isolée à partir d'une souche de Bt. Elle a un poids moléculaire de 23 kDa.	Ce sont des facteurs de virulence non - spécifiques	Taguchi et al., 1980; Agaisse et al., 1999
Activité hémolytique et entérotoxine			Agaisse et al., 1999; Siegel, 2001

Beta-exotoxine	701 Da	Il a été interdit pour usage du public, conformément aux recommandations de l'Organisation Mondiale de la Santé (WHO)	WHO, 1999; Espinasse et al., 2003

Jusqu'à présent, le rôle de la protéase dans l'accroissement ou la diminution de l'activité insecticide des delta-endotoxines contre les insectes nuisibles n'a pas été confirmé. Il convient de mentionner à nouveau que, après ingestion par des insectes sensibles, les cristaux sont solubilisés par le milieu alcalin de l'intestin des insectes et les cristaux protéiques solubilisés (les pro-toxines pleine longueur et inactifs) sont transformés en fragments plus petits et entomotoxiques.

Deux sources d'activité protéolytique sont capables de traiter et/ou de dégrader les cristaux pro-toxines: (a) l'activité protéolytique (dans l'intestin moyen) hydrolyse les pro-toxines en toxines actives (Haider et al., 1986; Ogiwara et al., 1992; Tojo et al., 1986); et (b) au cours de la sporulation, Bt synthétise des enzymes protéolytiques (Brar et al., 2007), et cette activité protéolytique peut traiter et / ou dégrader les cristaux protéiques (Andrews et al., 1985, Carroll et al., 1989; Chestukhina et al., 1980; Nickerson et al., 1981; Kunitate et al., 1989; Donovan et al., 1997; Tan et Donovan, 2000).

En fait, ils peuvent dégrader les protéines insecticides. Toutefois, leur absence ou leur présence ne modifient pas l'activité biopesticide des cristaux protéiques de Bt qui sont spécifiquement entomopathogéniques contre les lépidoptères (Nickerson et al., 1981; Donovan et al., 1997; Tan et Donovan, 2000). Toutefois, dans certains cas, l'inhibition de la protéase alcaline pourrait augmenter l'activité insecticide de Bt (MacIntosh et al., 1990). Ainsi, le rôle de ces protéases dans l'activité insecticide doit être vérifié (Brar et al., 2007).

1.2.1.4.4. Inhibiteurs immunitaires contre Bt

Bt est très résistant face au système humoral de défense de l'hôte, en particulier pour contre les cécropines et les attacines, qui sont les principales classes de peptides antibactériens inductibles chez divers lépidoptères et diptères (Dalhammar et Steiner, 1984; Edlund et al. 1976; Fedhila et al ., 2002). Une métalloprotéase contenant du zinc sécrétée par Bt, appelée Inha ou InA, en particulier hydrolyze les cécropines et les attacines dans l'hémolymphe de *Hyalophora cecropia* in vitro (Dalhammar et Steiner, 1984; Edlund et al. 1976).

Bien que la dégradation des cécropines et des attacines par Inha puisse expliquer en partie le succès de Bt pour envahir l'hemolymphe, l'importance de cette protéase dans la virulence a été débattue (Dalhammar et Steiner, 1984; Lövgren et al., 1990; Siden et al. 1979). En effet, le rôle de l'Inha dans la résistance au système immunitaire humorale n'est pas compatible avec la durée de sa production. L'expression tardive de l'Inha n'est pas compatible avec la production de l'anti-peptide, qui est une première réaction de défense de l'hôte (Fedhila et al., 2002).

Un autre rôle putatif est suggéré par le fait que l'Inha purifiée a un effet létal après injection chez l'insecte hôte. Les symptômes associés à l'administration de l'Inha sont typiques de la toxémie et non d'une septicémie bactérienne (Lövgren et al., 1990; Siden et al., 1979). Ainsi, les rôles de ces inhibiteurs immunitaire comme un agent possible de synergie (ou non) ne sont pas encore bien définies.

1.2.1.4.5. Phospholipases C

Les phospholipases sont un groupe hétérogène d'enzymes qui sont capables d'hydrolyser un ou plusieurs liens ester des glycérophospholipides. Les actions des phospholipases peuvent entraîner la déstabilisation des membranes, la lyse cellulaire et la libération de seconds messagers lipidiques. Ces enzymes sont classées en fonction de la spécificité de l'ester lien qui est clivé lors de leur action (Cox et al., 2001) (Figure 5).

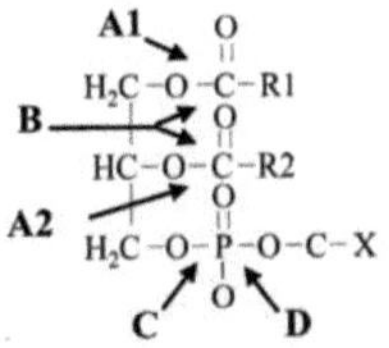

Figure 5. Sites de l'action de la phospholipase A1, A2, B, C et D

La structure générique de la phospholipase est représentée avec les sites d'action des phospholipases désigné par une flèche.

Les phospholipases C sont une composante des protéines actives qui peuvent dégrader la membrane (Elles comprennent la phospholipase C, la sphingomyélinase, la phospholipase C spécifique au phosphatidylinositol). (Gilmore et al., 1989). Une phospholipase C spécifique au phosphatidylinositol a été purifiée à partir du bouillon de culture de Bt à l'état homogéne comme indiqué par électrophorèse sur gel de polyacrylamide. L'enzyme purifiée (poids moléculaire: 23 000 + / - 1000) a une activité au maximale à un pH de 7.5 et n'est pas influencée par l'EDTA. L'enzyme spécifiquement hydrolysée agit sur le phosphatidylinositol, mais n'a aucune action contre la phosphatidylcholine, la phosphatidyléthanolamine, la phosphatidylsérine, le phosphatidylglycérol et la sphingomyéline (Taguchi et al., 1980). L'action de la phospholipase C agit contre les invertébrés et contre les vertébrés Sa présence est interdite dans les biopesticides à base de Bt (Agaisse et al., 1999).

1.2.1.4.6. Beta-exotoxine (β-exotoxine)

La β-exotoxine (Figure 6), également connu sous le nom de thuringiensine ($C_{22}H_{32}N_5O_{19}1P_1.3H_2O$), est un composé non-spécifique et stable à la chaleur. Elle est sécrétée dans le milieu de culture au début de la sporulation, par certaines souches de Bt. Ceci est un analogue insecticide de l'adénine-nucléotide, composé de l'adénosine, du glucose et de l'acide allarique et son poids moléculaire est de 701 Da (Bond et al., 1969). La toxicité de la β-exotoxine résulte probablement de l'inhibition de l'ARN polymérase et qui par concurrence avec l'ATP, inhibe la synthèse de l'ARN (Campell et al., 1987). Sa toxicité sur la mue et la nymphose et, à doses sublétales, ses effets tératologiques sur les différents ordres d'insectes a été décrite. En raison de sa toxicité pour les vertébrés, la plupart des préparations commerciales de Bt sont préparés à partir d'isolats qui ne produisent pas de β-exotoxine (Espinasse et al., 2003).

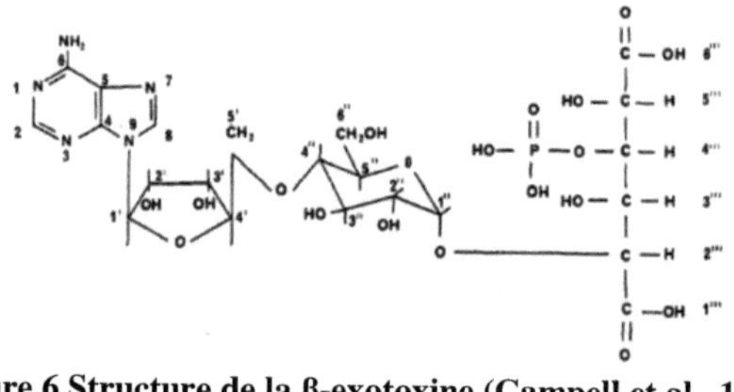

Figure 6.Structure de la β-exotoxine (Campell et al., 1987)

1.2.1.4.7. Zwittermicine A

Zwittermicine A est un antibiotique aminopolyol fongistatique de 396 Da (Figure 7) qui est produit par de nombreux bacilles, y compris les souches de Bt. Elle est produite au cours de la sporulation de Bt (Nair et al., 2004; Silo-Suh et al., 1998; Stabb et al., 1994; Stohl et al., 1999; Manker, 2002).

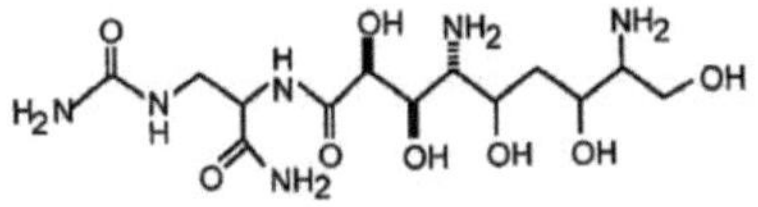

Figure 7. Structure de Zwittermicine A

Zwittermicine A obtenu à partir de *Bacillus cereus* peut augmenter l'activité insecticide de Bt subsp. *kurstaki*, alorsque seule, elle ne montre aucune activité insecticide lors d'un essai contre la spongieuse (*Lymantria dispar*) (Broderick et al., 2000, 2003). Il existe plusieurs mécanismes possibles par lesquels Zwittermicine A pourrait renforcer l'activité de Bt contre la spongieuse tels que (1) Action directe contre les cellules d'insectes, une fois qu'elles deviennent accessibles par le biais de la perforation de l'intestin moyen de l'épithélium de Bt; (2) Action directe sur divers aspects de la fonction de l'intestin moyen, telles que rupture de la membrane péritrophique ce qui supprime une barrière physique, la stimulation des protéases nécessaires pour solubiliser et activer des delta-endotoxines de Bt et la modification des propriétés de l'intestin moyen de l'épithélium pour faciliter la liaison des toxines de Bt et la formation de pores; (3) Les propriétés antimicrobiennes de Zwittermicine A peuvent modifier la composition de la microflore intestinale dans la spongieuse. La microflore intestinale est essentielle pour de nombreux insectes pour leur croissance normale, le développement, la reproduction, la digestion et la nutrition; La perturbation de ces relations pouvant potentiellement modifier la puissance de Bt (Broderick et al., 2000).

1.2.1.4.8. Résumé des caractéristiques de Bt var. kurstaki HD-1 (Btk HD-1)

La souche HD-1 de Bt var. kurstaki (Btk HD-1) est l'une des souches les plus efficaces et les plus utilisées dans de nombreux types de biopesticides commerciaux, car elle a un spectre d'activité assez large contre certains insectes ravageurs (Dulmage, 1970; Höfte et Whiteley, 1989). Cette souche a été utilisée dans notre laboratoire pour développer des processus de production économique des biopesticides. Le tableau 4 présente quelques caractéristiques de Btk HD-1.

Tableau 4. Quelques caractéristiques de Btk HD-1

Caractéristiques	Positive (+) or négative (-)	Rôle	Référence
Sérotype	Type: H3a3b (in 1973) and H3a3b3c (in 1999)	Classification	de Barjac and Bonnefoi, 1973; Lecadet et al., 1999
Séro-variété (serovar.)	*kurstaki*		de Barjac and Bonnefoi, 1973
Souche	HD-1		Dulmage, 1970
gènes codant pour Delta-endotoxines	*cry1Aa, cry1Ab, cry1Ac*	Synthèse des protéines de Cry1Aa, Cry1Ab, Cry1Ac (130-140 kDa)	Höfte and Whiteley, 1989
	cry2Aa and cry2Ab	Synthèse des protéines de Cry2Aa (activité contre à la fois Lépidoptères et Diptères), les	

		protéines de Cry2Ab protéines (activité contre Lépidoptère) (70 kDa)	
VIP 3A	+	VIP 3A est toxique contre une large gamme de Lépidoptère	Guttmann and Ellar, 2000; Donovan et al., 2001
Chitinase	+	Accroît l'activité biopesticide des delta-endotoxines de Bt	Guttmann and Ellar, 2000
Protéases alcalines	+	Mal défini	Tyagi et al., 2002
Métallo-protéases	+		Yong and Yousten, 1975
Alpha-exotoxine (phospholipase C or Lecithinase C)	+	Toxique contre un large éventail d'invertébrés et de vertébrés	de Barjac and Bonnefoi, 1973; Guttmann and Ellar, 2000
Inhibiteurs immunitaire A	+	Active contre le système immunitaire humoral des	Guttmann and Ellar, 2000

		insectes (cécropines et attacines)	
Activité hémolytique (hémolyse des érythrocytes humains sur géloses)	-	Toxique contre un large éventail d'invertébrés et de vertébrés	Carlson C.R., and Kolsto, 1993, Corlson et al., 1994
	+		Guttmann and Ellar, 2000; Hansen and Hendriksen, 2001
Entérotoxine (mesurée par immunoréaction)	+		Carlson C.R., and Kolsto, 1993, Corlson et al., 1994; Guttmann and Ellar, 2000
Beta-exotoxine type I and type II	-		Levinson et al., 1990
Zwittermicin A	+	Potentialiser l'activité biopesticide des delta-endotoxines de Bt	Manker et al., 1994; Stabb et al., 1994
Coproporphyrine	+	sensibilité de la spore à la lumière visible	Harms et al., 1986

1.2.2. Stratégies générales pour la production à haut rendement de δ-endotoxine et de biopesticides à base de Bt ayant un fort potentiel entomotoxique

La composante la plus importante pour tous les types de biopesticides à base de Bt est la concentration en delta-endotoxines (Schnepf et al., 1998). En pratique, la plupart des produits commerciaux de Bt sont basés sur le complexe spores et cristaux protéiques insecticides (ICPs), obtenu par centrifugation du bouillon de fermentation (Dulmage et al., 1970; Brar et al., 2006a). Pour améliorer l'efficacité de tous ces types de biopesticides, certaines stratégies et méthodes peuvent être appliquées. Afin d'évaluer l'effet de l'application de ces procédés, quelques recherches se sont basées sur le rendement de la production de delta-endotoxines par fermentation de Bt (Zouari et Jaoua, 1999); quelques travaux se sont basés sur le rendement des spores et ICPs ainsi que sur l'activité insecticide des produits fermentés (Morris et al., 1996). Enfin d'autres recherches ont été basées sur l'activité insecticide (entomotoxicité) des bouillons de fermentation de Bt (Brar et al., 2005a).

1.2.2.1. Optimisation des teneurs en éléments nutritifs pour la production de biopesticides à base de Bt

Dans le cas de l'utilisation de milieux semi-synthétiques pour une production optimale de biopesticides à base de Bt, les taux des éléments nutritifs ont été étudiés en tenant compte soit des ingrédients du milieu, comme par exemple la concentration en glucose (Scherrer et al., 1973; Arcas et al., 1987, Goldberg et al., 1980), soit sur le ratio optimal de C: N à différentes teneurs totales initiales de carbone (Farrera et al., 1998). Scherrer et al. (1973) ont constaté que, dans le milieu semi-synthétique pour la production de biopesticides à base de Bt, l'augmentation des concentrations de glucose permet d'obtenir des inclusions cristallines plus grandes avec une teneur élevée en protéines et en activité insecticide. Un rendement maximal en protéines et endotoxines est obtenu dans un milieu semi-synthétique contenant du glucose à une concentration de 6 à 8 g / L (Scheffer et al., 1973). Arcas et al. (1987) ont utilisé un milieu contenant du glucose et de l'extrait de levure pour la production de biopesticides. Ils ont constaté que la concentration de spores a été augmentée de 1,08 x 10^{12} spores/L à 7,36 x 10^{12} spores/L et les taux de toxines ont sont accrus de 1,05 mg/ml à 6,85 mg/ml lorsque la concentration de glucose a été portée de 8 à 56 g/L, avec augmentation du reste des composants du milieu : Une concentration plus élevée de nutriments réduisant la concentration de spores ou la production de toxines.

En ce qui concerne l'utilisation d'un milieu complexe tel que les boues ou autres déchets pour la production de biopesticides à base de Bt, il est bien connu que la plupart des éléments nutritifs nécessaires à la croissance Bt sont présents dans les solides, ce qui indique que l'augmentation de la concentration des solides dans le milieu peut accroître les taux de carbone et des autres nutriments. Ce fait a été prouvé dans le cas des boues d'épuration. L'optimisation des éléments nutritifs pour la production de biopesticides à base de Bt a été étudiée indirectement par rapport à la concentration de solides des boues et/ou le ratio de carbone: azote, ainsi que vis-à-vis les types de boues (Lachhab et al., 2001; Vidyarthi et al., 2002). Il a été constaté que la concentration de solides des boues influence fortement la croissance de Bt et l'entomotoxicité obtenue. Toutefois, une concentration de solides de 46 g/L réduit la concentration de cellules de Bt (5,4 × 10^8 unité formant colonie/ml ou UFC/ml), la concentration de spores (4,8 × 10^8 UFC / ml) et une faible entomotoxicité (9743 UI/μl) est obtenue; Le taux de sporulation étant de 89%. Le même phénomène a été observé avec une plus faible concentration de solides dans les boues (10 g/L) , entraînant une faible concentration de cellules ($4,3 \times 10^8$ UFC / ml), un faible taux de spores ($3,9 \times 10^8$ UFC/ml) et une entomotoxicité réduite (8231 UI/ µl). La concentration optimale des solides a été établie à 26 g/L, ce qui améliore l'entomotoxicité à 12970 UI/ul, les concentrations des cellules et de spores de $5,0 \times 10^9$ et $4,8 \times 10^9$ UFC/ml, respectivement, et un taux de sporulation de 96% (Lachhab et al., 2001). Par rapport aux taux de carbone et d'azote, il a été constaté qu'un faible rapport C/N dans les boues secondaires et un taux élevé de C / N dans les boues mélangées permettent d'obtenir une plus grande entomotoxicité; La valeur optimale de C: N dans les boues mélangées pour la production Bt ayant été établie à 7,9-9,9.

1.2.2.2. Ajouts de différentes sources de carbone et d'azote dans le milieu complexe pour augmenter l'entomotoxicité obtenue dans le bouillon de fermentation de Bt

Les effets de l'addition de glucose (comme source de carbone), extrait de levure (source d'azote organique) et de sulfate d'ammonium (source d'azote inorganique) sur la croissance de Bt et l'entomomotoxicité obtenue dans le bouillon de fermentation du Bt en utilisant des boues (non hydrolysées et hydrolysées) comme substrats ont été étudiés par LeBlanc (2004). Les résultats ont montré que l'apport de glucose (2 g/L) comme source de carbone dans des boues non-hydrolysées (Communauté-Urbaine-de-Québec - CUQ) a permis d'augmenter le potentiel d'entomotoxicité de biopesticide (de 10566 UI/µL jusqu'à 12376 UI/µl). Pour les boues hydrolysées, une faible augmentation de l'entomotoxicité a été observée par l'ajout de 2 g/L de glucose (de 14113 UI/µL à 15147 UI/µl). En outre, il a été constaté que le pourcentage élevé de la sporulation ou de haute concentration de spores ne permet pas d'accroître l'entomotoxicité (LeBlanc, 2004).

Un supplément d'extrait de levure (comme source d'azote) dans les boues n'a pas entraîné de différence significative au point de vue de la concentration de cellules de Bt. Toutefois, dans les boues non hydrolysées, la concentration de spores a été réduite suite à une augmentation de la concentration d'extrait de levure ajouté à la boue. En ce qui concerne l'entomotoxicité, l'addition de 1 ou de 2 g/L d'extrait de levure dans les boues non hydrolysées permet de l'augmenter de 10566 UI/µL (contrôle) à 12912 et 13300 UI/µL, respectivement. Dans les boues hydrolysées, une légère augmentation de l'entomotoxicité a été observée (de 14113 à 15406 UI /µl) suite à l'addition de 1 g/L d'extrait de levure (LeBlanc, 2004). D'autre part, les résultats obtenus au cours des expériences réalisées montre que l'ajout de sulfate d'ammoniac comme source d'azote inorganique dans un mélange de boues non hydrolysées et hydrolysées n'a pas accru l'entomotoxicité. Il est possible que les boues aient assez de source d'azote inorganique (LeBlanc, 2004).

1.2.2.3. Addition de NaCl et de Tween 60 ou Tween 80 dans le milieu de fermentation

L'addition de 0,5% de NaCl et de 0,1% de Tween 60 à un milieu de culture contenant des pulpes de graines de coton (Proflo) et de glucose comme les principales sources d'hydrates de carbone et d'azote, respectivement, a augmenté l'activité insecticide (spores- cristaux) de Bt var. *aizawai* (HD133) (spores- cristaux) Morris et al., 1996). Dans le cas de l'utilisation de gruau brut et de farine de poisson comme milieu complexe pour la production de biopesticides de Bt var. *kurstaki* BNS3, la présence de 0,5 g/L de NaCl et de 0,1% de Tween-80 accroît le rendement en delta-endotoxines (Zouari et Jaoua, 1999).

Récemment, l'effet d'un agent tensio-actif, le Tween 80 (0,2%, v/v) sur la production de biopesticides à base de Bt en utilisant des boues d'épuration secondaires (non hydrolysées (NH) et hydrolysées (TH)) comme substrats de fermentation a été étudié en bioréacteurs. Boues hydrolysées ont permis d'augmenter l'entomotoxicité (Tx) (+49%) et la concentration de cellules et spores vis-à-vis ce qui est produit à l'aide des boues non hydrolysées. L'ajout de Tween 80 à des boues non hydrolysées a augmenté de 1,67 et 4 fois respectivement, les teneurs en cellules et en spores; Le taux maximum de croissance spécifique (µmax) ayant augmenté de 0,19 à 0,24 h^{-1} et celui de l'entomotoxicité (Tx) de 26,6%. Toutefois, l'apport de Tween 80 à des boues hydrolysées a augmenté de 2 et 2,4 respectivement les taux de cellules et de spores, et la croissance spécifique µmax est passée de 0,28 à 0,3 h^{-1} sans modification de l'entomotoxicité (Tx) (Brar et al., 2005a). Ces résultats intéressants montrent que l'ajout de NaCl et de Tween 60 ou de NaCl et de Tween 80 ou de Tween 80 dans le milieu de fermentation augmente considérablement le rendement en puissance insecticide et en production de delta-endotoxines et ce, indépendamment des types de milieu (à l'exception de boues hydrolysées).

Ceci peut être expliqué par le fait que de faibles concentrations de sels neutres comme NaCl augmente la solubilité de nombreuses protéines alors que le Tween 80 (polyoxyéthylène sorbitane mono-oléate) et des agents tensio-actifs ont été utilisés en cultures bactériennes afin de promouvoir l'adsorption de composés dans les cellules tout en augmentant la disponibilité des protéines solubles dans le milieu. Ainsi, l'addition de ces suppléments dans le milieu de culture augmente probablement la solubilité et de la disponibilité pour Bt, des protéines lors de la production des sous-unités entomotoxiques de l'endotoxine (Morris et al., 1996; Zouari et Jaoua, 1999; Brar et al., 2005a).

1.2.2.4. Mode de culture Fed batch (Cultures discontinues avec un apport contrôlé de substrat)

Avec un milieu synthétique classique pour l'obtention de biopesticides à base de Bt ayant une haute densité de cellules et de spores, le mode de culture fed-batch est susceptible d'être la meilleure stratégie. Kang et al. (1992) a appliqué un fed-batch intermittent pour augmenter la production de cellules et de spores de Bt, celui-ci a atteint le nombre final de 12,53 x 10^9 spores/mL. Jong et al. (1994) ont évalué une culture fed-batch de Bt fondée sur le contrôle du pH et ont rapporté un nombre maximum de spores de 8,3 x 10^9 /ml. En employant une stratégie de fed-batch basée sur une forte agitation, Chen et al. (2003) ont montré que la concentration de cellules a augmenté de 50% par rapport à la culture batch. Cependant, toutes ces études ont été effectuées à l'aide d'un milieu synthétique et la valeur d'entomotoxicité n'a pas été rapportée.

Récemment, une stratégie simple de fed-batch basée sur la mesure de l'oxygène dissous (DO) au cours de la fermentation a été développée en utilisant des boues d'épuration comme substrat pour produire des biopesticides à base de Bt. Il a été constaté par le remplacement de la stratégie du processus de batch par fed-batch, que la concentration de spores a été augmentée de 5,62 x 10^8 à 8,6 x 10^8 (UFC/ml), ce qui a abouti à élever l'entomotoxicité de 13 x 10^9 à 18 x 10^9 SBU/L. Une entomotoxicité supérieure a été enregistrée avec une faible concentration de spores en utilisant des boues d'épuration comme substrat alors qu'une entomotoxicité faible a été notée avec une haute concentration de spores dans le milieu synthétique (Yezza et al., 2005a).

1.2.2.5. Utilisation de différents agents de contrôle du pH au cours de la fermentation

L'hydroxyde d'ammonium et l'acide acétique (NH_4OH/CH_3COOH) ont été utilisés comme agents de contrôle du pH au cours de l'obtention de Bt dans un fermenteur de l'échelle pilote (150-l) en employant deux boues secondaires provenant de deux usines de traitement des eaux usées (CUQS et JQS) et un milieu semi-synthétique à base de tourteau de soja comme substrats (Yezza et al., 2005c). Les résultats ont montré une forte augmentation des concentrations totales des cellules, des spores, de l'activité des protéases et de l'entomotoxicité lorsque le pH de la culture a été contrôlé en utilisant NH_4OH/CH_3COOH et ce, indépendamment du milieu. À la fin de la fermentation (48h), la concentration totale de cellules a augmenté de près de 17%, 33% et 25%, l'activité des protéases a cru de 12%, 33% et 53% et la concentration maximale de spores a augmenté de près de 28%, 48% et 33% dans les milieux de CUQS, JQS et le soja, respectivement. L'entomotoxicité s'est élevée de 22%, 21% et 14% dans les milieux de CUQS, JQS et le soja, respectivement, par rapport aux résultats obtenus avec $NaOH/H_2SO_4$ pour contrôler le pH.

Une plus grande entomotoxicité a également été observée en utilisant des boues par rapport à ce qui est observé avec un milieu semi-synthétique. Cette amélioration de la performance du processus à base de Bt est une conséquence d'un ajout rapide de la source de carbone et d'azote utilisable par le biais de contrôle du pH, ce qui a stimulé la production d'endotoxine et a augmenté la sporulation (Yezza et al., 2005c).

1.2.2.6. Traitement des boues d'épuration pour augmenter la biodégradabilité des nutriments

Les boues d'épuration sont des matières complexes pour la fermentation et le prétraitement est nécessaire pour transformer les composés moins biodégradables en nutriments plus facilement accessibles (Barnabé, 2004, Yezza et al., 2005b). En utilisant de nombreuses méthodes telles que l'hydrolyse acide, alcaline ou thermique, ainsi que l'oxydation par H_2O_2 pour le traitement des boues mixtes et secondaires, il a été constaté que l'emploi de boues hydrolysées par traitement alcalin, alcalino-thermique, ou thermo-oxydant permet d'améliorer significativement l'entomotoxicité des biopesticides à base de Bt (Barnabé, 2004). Les conditions optimales de l'hydrolyse alcaline, thermo-alcaline, et thermo-oxydante ont été établies en fermenteur de 15-L. Les entomotoxicités obtenues variaient entre 17400 et 19000 UI/μL avec les boues traitées (solides en suspension de 37,4 g/L) soit de 37% à 49% de plus que ce qui est observé avec les boues non traitées (solides en suspension à de 25,0 g/L) et de 1,7 à 1,9 fois plus élevé que ce qui est constaté avec le milieu contenant du soja conventionnel (semi-synthétique).

D'autres résultats intéressants ont été obtenus par Yezza et al. (2005b). Dans cette étude, les boues ont subi un traitement thermo- alcalin ou une oxydation et employées pour produire Bt dans fermenteurs de 15-L dans des conditions contrôlées. Il a été constaté que l'hydrolyse thermo- alcaline est un processus efficace afin d'accroître l'entomotoxicité de Bt contre la tordeuse des bourgeons de l'épinette. La concentration des cellules et des spores s'est élevée de 40 et 46% respectivement par rapport à celle rapportée avec des boues non traitées. L'entomotoxicité finale augmente de 12,3 $x10^9$ SBU/L avec des boues brutes à 16,6 x 10^9 SBU/L avec la boue traitée par des agents thermo-alcalins. L'amélioration de la performance du processus a été attribuée à un meilleur transfert d'oxygène résultant d'une diminution de la viscosité des milieux et d'une plus grande disponibilité des éléments nutritifs en raison de la solubilisation et de la biodégradabilité des boues (Barnabé, 2004; Yezza et al.2005b).

1.2.2.7. Chitinases endogènes comme agents de synergie pour augmenter l'entomotoxicité des biopesticides à base de Bt

Wiwat et al., 2000 ont isolé une souche de *B. thuringiensis* ssp. *kurstaki* HD-1 (G) qui a été identifiée comme celle produisant le plus de chitinase dans leur étude. Ce Bt synthétise la plus grande quantité de chitinase (19,3 mU/ml) quand il est cultivé sur un agitateur rotatif (200 rpm) à 30°C pendant 2 jours, dans un bouillon de nutriments contenant 0,3% de chitine colloïdale. L'entomotoxicité de *B. thuringiensis* ssp. *kurstaki* HD-1 (G) et celle de *B. thuringiensis* ssp. *kurstaki* wa-p-2, (non producteur de la chitinase) ont été comparées vis-à-vis des larves de *Plutella xylostella* (fausse-teigne des crucifères) Ils ont noté que les concentrations létales 50% (CL_{50}) étaient respectivement de 4,93 x 10^4 et 1,32 x 10^5 spores/ml, (plus la valeur de la CL 50% est faible, plus la toxicité est élevée). Quand le bouillon de culture de *B. thuringiensis* ssp. *kurstaki* HD-1 (G) a été utilisé comme liquide de suspension au lieu d'un tampon phosphate, les valeurs des CL 50 ont été réduites à 6,23 x 10^3 et 7,60 x 10^4 spores/ml, respectivement. Leurs études ont indiqué que la toxicité de *B. thuringiensis* ssp. *kurstaki* HD-1 (G) pour les larves de fausse-teigne des crucifères augmente en cas d'utilisation des surnageants.

Ces travaux montrent que la chitinase présente dans le surnageant et/ou certains facteurs peutvent accroître significativement la toxicité de *B. thuringiensis* ssp. *kurstaki* HD-1 (G). Par conséquent, l'amélioration de la toxicité des biopesticides à base de *B. thuringiensis* par addition de surnageant de culture dans le réservoir de mélange avant l'application sur le terrain peut être envisagée (Wiwat et al., 2000). Cette recherche est très intéressante du point de vue de l'application pratique. Toutefois, elle a été menée uniquement sur milieu synthétique. En outre, ces travaux n'ont utilisé que la concentration de spores pour évaluer la toxicité.

1.2.2.8. Zwittermicine A comme agent de synergie pour augmenter l'entomotoxicité des biopesticides-Bt

Jusqu'à présent, une seule recherche a examiné l'addition des Zwittermicines A produites par *Bacillus cereus* comme agent de synergie pour accroître l'entomotoxicité du biopesticide commercial 'Foray76B' (Broderick et al., 2000). Dans cette étude, les auteurs ont constaté que *B. cereus* UW85 produit la Zwittermicine A à forte concentration dans un milieu tryptique soja et ils ont démontré que l'ajout d'une culture de *B. cereus* UW85 (sur un agitateur rotatif (200 rpm) à 30 ° C pendant 2 jours) augmente de façon significative le taux de mortalité de larves de spongieuse. Par ailleurs lorsqu'il est appliqué seul (3,3 %), *B. cereus* UW85 n'a eu aucun effet sur la survie des larves de spongieuse. Une faible concentration de culture d'UW85 (0,033 µl par disque) a presque doublé l'activité de *B. thuringiensis* subsp. *kurstaki* contre les larves, alors que *B. thuringiensis* subsp. *kurstaki* seul, a causé seulement

20% de mortalité. Une augmentation de la mortalité des spongieuses a été notée avec des doses plus élevées, atteignant 95%, avec un apport de 10μl de la culture. Les auteurs ont également relevé que la Zwittermicine A a un effet significatif sur le temps de mortalité des larves provoquée par Bt. Au 3ième jour des bio-essais, *B.thuringiensis* seul a causé la mortalité de seulement 17% des chenilles. Avec l'addition de zwittermicine A à Bt, la mortalité larvaire variait de 61% avec un apport de 500 pg à 93 % avec un ajout 150 ng de Zwittermicine. Ces travaux de R&D peuvent conduire à développer une méthode avancée pour la production de haute activité biopesticide à base de Bt ayant un potentiel accru. Toutefois, cette étude a utilisé *B. cereus* pour la production de Zwittermicine A. Elle ne donnait aucune information sur la production de cet antibiotique par Btk HD-1 et de la récupération de la Zwittermicine A pour la formulation de biopesticide -Bt.

1.2.3 Problématiques

Les recherches antérieures sur l'utilisation les eaux usées d'industrie d'amidon (SIW) comme matières premières pour la production de biopesticides à base de Bt ont été réalisées avec les SIW de la compagnie ADM-Ogilvie à Candiac (QC, Canada). Ces eaux usées ont une concentration de solides totaux de 15 à 17 g/L (Brar et al., 2005b; Yezza et al., 2007). Ces recherches ont démontré que les SIW comportaient certains avantages par rapport à l'utilisation des boues : (1) elles peuvent être utilisées directement comme matières premières sans traitement pour la fermentation (Brar et al., 2005b; Yezza et al., 2006); (2) elles sont appropriées pour la fermentation et ce, sans problème de formation de mousse (Brar et al., 2005b; Yezza et al., 2007); (3) les biopesticides produits à base de Bt en utilisant les SIW comme matières premières sont aussi valables pour la réalisation de formulations finales (Brar et al., 2005b). Toutefois, ces recherches sont encore préliminaires et il reste de nombreux problèmes à résoudre:

a. La delta-endotoxine est la principale composante des biopesticides à base de Bt pour le contrôle biologique des insectes ravageurs ciblés. Toutefois, la concentration en delta-endotoxines produites par Btk HD-1 dans les différents substrats de fermentation (SIW ou boues d'épuration ou milieu semi-synthétique) n'a pas été déterminée;

b. En pratique, la plupart des produits de Bt commerciaux sont basés sur le complexe de spores et de cristaux protéines insecticides (ICPs) qui peut être obtenu par centrifugation du bouillon de fermentation. Ce complexe de spores et des ICPs (culot de centrifugation) peut être remis en suspension avec le surnageant du bouillon de fermentation pour obtenir un nouveau mélange qui est ensuite utilisé pour formulation et obtention du produit final. Par conséquent, pour évaluer l'efficacité des biopesticides produits par tous les procédés de fermentation, il est nécessaire de déterminer l'entomotoxicité (Tx) du mélange par des bioessais. Cela n'a pas étudié dans les recherches antérieures;

c. La concentration en solides totaux des SIW (~ 15 à 17 g/L) n'est pas aussi élevée que celle en solides totaux des boues (~ 30 g / L) ou des milieux semi-synthétiques à base de la farine de soja (~ 28 g / L) qui sont normalement utilisés pour la production de biopesticides dans des recherches antérieures. Il n'est pas possible pour l'instant de savoir si les SIW ont suffisamment de nutriments pour la production de biopesticides à base de Bt. Par conséquent, une question est soulevée : « Est-il possible d'appliquer certaines méthodes connues pour améliorer la Tx en modifiant la composition en nutriments dans les SIW comme l'optimisation de la concentration en solides totaux du milieu; la fortification en nutriments du milieu au cours de la fermentation en ajoutant différents agents pour le contrôle du pH; l'enrichissement en nutriments du milieu avant la fermentation avec des sources adéquates de carbone ou azote; et l'application du mode de culture fed- batch, batch fermentation)? ».

d. Des méthodes avancées pour l'amélioration de Tx basée sur la synergie d'action des chitinases ou de la Zwittermicine A (qui peut être produite par Btk HD-1- Tableau 4) avec les delta-endotoxines et les spores de biopesticides n'ont pas été explorées;

e. La réponse à la question suivante n'est pas encore connue : « Est-il possible d'établir les relations entre les concentrations en spores et en delta-endotoxines, entre la concentration de spores et de Tx, et entre la concentration des delta-endotoxines et Tx? ».

1.3. OBJECTIFS – HYPOTHÈSES – ORIGINALITÉ DE LA RECHERCHE

1.3.1. Objectifs de recherche

1.3.1.1. Objectif global

Production de biopesticides à base de Btk HD-1 avec une très forte activité en utilisant les SIW comme matières premières.

1.3.1.2. Objectifs spécifiques

a. Addition de nutriments supplémentaires par l'utilisation de différents agents de contrôle du pH pour augmenter l'activité pesticide de Btk HD-1 en employant les SIW comme matières premières.

b. Optimisation de concentration en solides des SIW pour l'obtention de hauts rendements en delta-endotoxine et en Tx de Btk HD-1.

c. Augmentation de la production en delta-endotoxine et de Tx des biopesticides à base de Btk HD-1 par l'ajout de sources appropriées de carbone et/ou d'azote ou de Tween 80 dans les SIW utilisées comme matières premières.

d. Fermentation en fed-batch afin d'augmenter la production de delta-entotoxines de Btk HD-1 en utilisant les SIW comme matières premières.

e. Production induite des chitinases pour améliorer la Tx de *Bacillus thuringiensis* avec les SIW comme substrat de fermentation.

f. Récupération des chitinases et Zwittermicine A à partir de surnageant de fermentation de *Bacillus thuringiensis* pour produire des biopesticides avec une très forte activité insecticide.

g. Relations entre la concentration de delta-endotoxines et celle des spores, entre les concentrations de spores et Tx ou entre la concentration de delta-endotoxines et Tx de différents bouillons fermentés (milieu semi-synthétique, boues secondaire et SIW).

1.3.2. Hypothèses de recherche

1.3.2.1. Hypothèse générale

SIW ne pas avoir suffisamment de nutriments pour maximiser la production de biopesticide à base de Btk HD-1 avec une haute entomotoxicité et il est possible d'appliquer certaines méthodes pour la production de biopesticide à base de Btk HD-1 avec un potentiel élevé en utilisant les SIW comme matières premières.

1.3.2.2. Hypothèses spécifiques

a. Durant la fermentation de Btk HD-1 en utilisant les SIW comme matières premières, le pH peut être contrôlé (à 7) en employant 4 combinaisons d'acides et bases conventionnels ou non conventionnels : NaOH et H_2SO_4 (agents conventionnels); NH_4OH et CH_3COOH; NH_4OH et H_2SO_4; et NaOH et CH_3COOH. **L'utilisation de ces agents de contrôle du pH occasionnerait un apport supplémentaire en ions (H^+ et OH^-) et/ou en**

minéraux et nutriments (Na^+, NH_4^+, CH_3COO^-, SO_4^{2-}) dans le milieu de fermentation. Il est possible que certains de ces éléments puissent supporter ou stimuler la croissance, la production d'enzymes et finalement le rendement en Tx du bouillon fermenté de Btk HD-1. Ainsi, des expériences sont requises pour choisir un ou les combinaisons capables (s) d'accroître la tx.

b. Les SIW peuvent être divisées en deux fractions : la fraction soluble (contenant des protéines solubles, des sucres simples, des minéraux, etc.) et la fraction solide (contenant des matières insolubles dont des protéines complexes, des polysaccharides, des fibres, etc.). **Diverses concentrations en solides de SIW (obtenues en mélangeant la fraction solide avec la fraction soluble selon diverses proportions) lors de son emploi pour la production de Btk HD-1 auraient différents effets sur la croissance, la synthèse des delta-endotoxines et finalement la Tx de Btk HD-1.** Des expériences sont requises pour déterminer la concentration optimale en solides de SIW capable d'augmenter l'entomotoxicité de Btk HD-1 par rapport à l'emploi de SIW sans l'ajustement préalable de la concentration en solides.

c. **Avec l'ajout de différentes sources supplémentaires de carbone et d'azote (à différentes concentrations) ou de Tween 80 (à diverses concentrations)** dans les SIW pour la production de Btk HD-1, les milieux modifiés ont-ils différents effets sur la croissance, la synthèse des delta-endotoxines et finalement la Tx du bouillon fermenté de Btk HD-1. En réalisant des expériences sur ce sujet, il est possible de choisir les sources appropriées de carbone ou d'azote ou de Tween 80 et/ou et déterminer leur concentration idéale pour obtenir un biopesticide de Btk HD-1 avec une Tx plus élevée.

d. La fermentation en **fed-batch** en employant les SIW comme matières premières pour produire Btk HD-1 pourrait augmenter le rendement en delta-endotoxines et la Tx avec les paramètres adéquats (temps de fermentation auquel les ajouts sont faits; volume des additions; nombre d'alimentation; concentration en solides des alimentations). Les indicateurs indirects des nutriments disponibles dans le milieu durant la fermentation (comme le profil du pH ou pH-stat, et le profil de l'oxygène dissous ou DO-Stat) pourraient être employés pour optimiser ces paramètres. En étudiant et en optimisant les paramètres pour la fermentation en fed-batch, une Tx plus élevée pourrait être obtenue.

e. **Btk HD-1 a la capacité de produire des chitinases dans des milieux synthétiques à base de chitine où cette dernière est utilisée comme substrat induisant la production des enzymes en question.** L'addition de la chitine colloïdale à une concentration appropriée dans les SIW permettrait à Btk HD-1 de produire des chitinases durant la fermentation et qui seraient présentes dans le surnageant du bouillon fermenté après centrifugation. Il est possible que ces chitinases puissent augmenter la Tx du bouillon fermenté de Btk HD-1 en agissant en synergie avec d'autres facteurs de virulence. En outre, la récupération des chitinases du surnageant du bouillon de fermentation par ultrafiltration et l'utilisation de ces enzymes récupérées pour le mélange avec du culot de centrifugation (complexe de spores-cristaux de protéines) peut produire des biopesticides à haute Tx.

f. Il est connu que Btk HD-1 peut produire la **Zwittermicine A**, une molécule soluble susceptible d'accroître la Tx envers l'insecte ciblé. Btk HD-1 a le potentiel de produire la Zwittermicine A dans les SIW durant la fermentation. Il est possible également que la Zwittermicine A soit présente dans le surnageant du bouillon fermenté (après centrifugation) et qu'elle puisse être récupérée dans le filtrat suite à l'ultrafiltration du surnageant. Par conséquent, la récupération de la Zwittermicin A et son addition au culot de centrifugation contenant le complexe spores-protéines cristallines permettraient d'augmenter la Tx contre les larves de la tordeuse des bourgeons de l'épinette.

g. Il est finalement connu que **les delta-endotoxines sont parmi les plus importants composants des biopesticides à base de Bt ou de Btk HD-1.** Il est connu aussi que la réalisation des bio-essais est nécessaire pour identifier la Tx des biopesticides à base de Bt. Cependant, ces procédés sont laborieux et coûteux. Ainsi, est-il possible d'établir une relation entre la concentration en delta-endotoxine et la tx des biopesticides à base de Btk HD-1 et obtenus dans les milieux alternatifs tels que les boues et les SIW?

1.3.3. Originalité de recherche

Les recherches antérieures sur l'utilisation des SIW comme matières premières pour la production de biopesticides basés sur Btk HD-1 dans le cadre du projet Bt-INRS a démontré que les SIW ont quelques avantages par rapport à l'emploi de boues d'épuration et du milieu semi-synthétique dont une activité bioinsecticide plus élevée. Cependant, afin de bénéficier de tous les avantages de l'utilisation des SIW, il existe encore des lacunes à combler, clairement décrits dans la section 1.2.3. **Ainsi, les objectifs proposés sont originaux et n'ont jamais fait l'objet de recherches jusqu'à maintenant. Par conséquent, des recherches systématiques sont requises pour maximiser le potentiel pesticide et pour développer un procédé économique de production de biopesticides en utilisant les SIW comme matières premières selon les objectifs proposés.**

À l'exception de l'optimisation des paramètres du procédé, un effort sera fait pour corréler la concentration en delta-endotoxines et en spores avec la Tx contre les larves de la tordeuse des bourgeons de l'épinette. Actuellement, aucune information n'est disponible sur ces aspects. Normalement, les travaux antérieurs de plusieurs chercheurs utilisaient la concentration en spores ou en toxines, ce qui pouvait ne pas avoir une relation directe avec la

Tx qui s'avère être le but ultime du procédé de production. **Par conséquent, la recherche proposée est originale et répondra à beaucoup de questions pratiques et théoriques sur plusieurs aspects de la production de Btk HD-1.**

1.4. METHODOLOGIE

1.4.1. Fortification des nutriments par différents agents de contrôle du pH pour augmenter l'activité biopesticide de Bt en utilisant des SIW comme substrat de fermentation

Au cours de la fermentation de Btk HD-1 en bioréacteur de 15 L en utilisant des SIW comme substrat de fermentation, 4 agents de contrôle du pH soit $NaOH/H_2SO_4$ (agents conventionnels de contrôle du pH), NH_4OH/CH_3COOH NH_4OH/H_2SO_4 et $NaOH/CH_3COOH$ ont été utilisés pour maintenir le pH du milieu à 7. Les quantités d'agents de contrôle du pH ont été enregistrées en terme d'ions H^+ et OH^- (mol) et de minéraux (nutriments), tels que Na^+, SO_4^{2-}, NH_4^+, CH_3COO^- (g/L) qui ont été ajoutés dans les SIW au cours de la fermentation. Des échantillons ont été retirés à intervalles réguliers aux cours de la fermentation et ont été analysés pour évaluer les effets de l'ajout d'éléments nutritifs sur les paramètres suivants: (a) croissance (cellules, des taux de croissance), (b) spores et taux de sporulation, (c) production d'enzymes (protéases alcalines, amylase), (d) Tx du bouillon de fermentation et Tx de la suspension des culots de centrifugation (bouillon de fermentation concentré 10 fois par centrifugation). Les résultats obtenus seront utilisés pour évaluer le pH des agents de contrôle qui peuvent améliorer la Tx (Vu et al., 2008).

1.4.2. Optimisation de la concentration des solides totaux de SIW pour la production de biopesticide à base de Btk HD-1

La décantation permis de fractionner les SIW en 2 parties: matières en solides concentrées et matières surnageantes. Deux fractions ont été mélangées à différents ratios pour obtenir diverses concentrations des solides totaux de SIW (15, 23, 30, 35, 43, 48, 55, 61 et 66 g/L). Ces solutions ont été utilisées comme milieux pour l'obtention de Bt en erlenmeyers dans un agitateur rotatif pendant 48h à 30°C et 220 tr/min en introduisant un volume d'inoculum de 2%. Les échantillons ont été prélevés après 48h de fermentation afin d'analyser les teneurs en cellules, spores et delta-endotoxines. La concentration des solides totaux de SIW dans laquelle la plus haute concentration de delta-endotoxines obtenue a été choisie pour procéder à d'autres séries d'expériences afin de déterminer le meilleur volume d'inoculum de pré-culture (%, v/v).

La production de Btk HD-1 en bioréacteur de 15-L a été réalisée avec la concentration optimale des solides totaux de SIW et le volume optimal d'inoculum de la pré-culture qui ont été sélectionnés par les experiences en erlenmeyers. En outre, le milieu semi-synthétique à base de farine de soja et SIW avec la concentration d'origine de solides totaux (SIW sans modification) ont également été utilisées pour la fermentation de Btk HD-1 dans des bioréacteurs comme contrôles. Des échantillons ont été retirés à partir des bioréacteurs à intervalles réguliers au cours de la fermentation et ont été utilisés pour analyser les items suivants: (a) paramètres de fermentation tels que coefficient volumétrique de transfert d'oxygène (k_La), taux de transfert de l'oxygène (OTR) et taux d'absorption d'oxygène (OUR); (b) croissance (concentration des cellules et taux de croissance), (c) concentration des spores et taux de sporulation; (d) production d'enzymes (protéases alcalines, amylases); (e) Tx du bouillon de fermentation, et (f) Tx de suspension du culot (fermenté bouillon concentré 10 fois par centrifugation). Les résultats obtenus seront utilisés pour évaluer les effets globaux de l'utilisation de la concentration de solides totaux optimale des SIW (Vu et al., 2009a).

1.4.3. Production de biopesticide par enrichissement par des sources de carbone et/ou de l'azote approprié dans les SIW comme matières premières

Différentes sources de carbone (glucose, amidon de maïs et Tween 80) et d'azote (extrait de levure, peptone, extrait de viande bovine, peptone de caséine) ont été ajoutées dans les SIW pour obtenir diverses concentrations finales. Tous les mélanges ont été utilisés comme matières premières pour l'obtention en erlenmeyers de Bt pendant 48h à 30°C et 220 tr/min en utilisant un volume d'inoculum de 2%. Des échantillons ont été prélevés des erlenmeyers après 48h de fermentation et utilisés pour analyser les concentrations en cellules, spores et delta-endotoxines. Des SIW fortifiés avec des source (s) de carbone ou d'azote à la concentration ayant donné la plus grande concentration de delta-endotoxines ont été choisis pour réaliser des expériences dans des bioréacteurs pour confirmer les résultats.

La production de Btk HD-1 en bioréacteur de 15 L a été effectuée avec des SIW fortifiées avec des sources de carbone ou d'azote (s) qui ont été sélectionnées par les tests en erlenmeyer. En outre, le milieu semi-synthétique (farine de soja) et les SIW (sans modification) ont également été utilisés comme contrôles pour la fermentation de Btk HD-1 dans des bioréacteurs. Des échantillons ont été prélevés des bioréacteurs à intervalles réguliers au cours de la fermentation et ont été soumis à l'analyse des données suivantes: (a) paramètres de fermentation tels que coefficient volumétrique de transfert d'oxygène (k_La), taux de transfert de l'oxygène (OTR) et taux d'absorption d'oxygène (OUR); (b) croissance (concentration en cellules et taux de croissance), (c) concentration en spores et taux de sporulation; (d) production d'enzymes (protéases alcalines, amylases); (e) Tx du bouillon de fermentation, et (f) Tx de suspension du culot (bouillon fermenté concentré 10 fois par centrifugation). Les résultats obtenus seront utilisés pour évaluer les effets globaux de l'utilisation de SIW fortifiée avec des source (s) de carbone ou l'azote (Vu et al., 2009b).

1.4.4. Mode de culture Fed-batch de *Bacillus thuringiensis* utilisant les SIW comme substrat de fermentation

Le mode de culture Fed-batch a été conduit en se basant sur le taux d'oxygène dissous (DO) comme paramètre de contrôle rétroactif et par utilisation du mode manuel de l'intermittence de l'alimentation par des éléments nutritifs (présents dans les SIW) dans le milieu de fermentation. Trois différents nutriments ont été ajoutés dans le fermenteur et le temps de l'alimentation a été déterminé en fonction de la DO mesurée lors du processus de fermentation.

L'opération a débuté en mode batch, avec un volume de travail de 8 L. Après inoculation, la concentration d'oxygène dissous a commencé à baisser jusqu'à la fin de la croissance exponentielle. Après cela, la DO a commencé à augmenter. Lorsque la DO approchait de la stabilisation, 2 L de SIW (avec de solides totaux de 30g/L) frais stérilisés ont été ajoutés dans le fermenteur. L'oxygène dissous a chuté à un minimum suivi par une augmentation. Lorsque la DO atteint près du maximum, 3L du contenu du fermenteur ont été retirés et remplacés par un même volume de SIW (avec de solides totaux de 30 g/L) frais stérilisés. La valeur de la DO à nouveau a diminué, et ensuite par une augmentation tout prêt du maximum, 3L du contenu du fermenteur ont été retirés et remplacés par le même volume de SIW (avec de solides totaux de 30 g/L) frais stérilisés. Le processus de fermentation a été poursuivi en mode batch jusqu'à 72 h. Des échantillons ont été prélevés du fermenteur à intervalles réguliers afin de déterminer les concentrations en cellules, spores, delta-endotoxines et valeur de la Tx. En outre, les échantillons qui ont obtenu les plus hautes valeurs de Tx à chaque exécution de fermentation ont été employés pour évaluer par bioessais les valeurs de la Tx de la suspension du culot (Vu et al., 2009c).

1.4.5. Production induite des chitinases pour améliorer la Tx de *Bacillus thuringiensis* en utilisant les SIW comme substrat de fermentation

Différentes concentrations de chitine colloïdale ont été ajoutées dans les SIW pour obtenir une concentration finale de 0,0 à 0,30%, poids/volume (p/v). Ces solutions ont été utilisées comme milieux pour la production de biopesticide à base de Btk HD-1 en erlenmeyer (à 30°C, agitation de 220 tr/min pendant 48h). Des échantillons ont été prélevés des erlenmeyers après 48h pour analyser les concentrations en cellules, en spores, et les activités des chitinases ainsi que des valeurs de la Tx contre des larves de la tordeuse des bourgeons de l'épinette. En outre, les valeurs de la Tx de la suspension du culot de ces échantillons ont également été estimées.

En se basant sur les résultats obtenus par fermentation en erlenmeyers, les SIW complétés avec de la chitine colloïdale à la concentration optimale qui améliore la Tx du biopesticide ont été employé pour procéder à de nouvelles expériences en bioréacteur (15-L) pour confirmer les résultats. Des échantillons ont été prélevés au cours de la fermentation, à intervalles réguliers afin d'analyser ce qui suit: (a) paramètres de fermentation tels que coefficient volumétrique de transfert d'oxygène (k_La), taux de transfert de l'oxygène (OTR) et taux d'absorption d'oxygène (OUR); (b) croissance (concentration en cellules et taux de croissance), (c) concentration en spores et taux de sporulation; (d) production d'enzymes (protéases alcalines, amylases); (e) Tx du bouillon de fermentation, et (f) Tx de suspension du culot (bouillon fermenté concentré 10 fois par centrifugation). Les résultats obtenus seront utilisés pour évaluer les effets globaux de l'utilisation de SIW fortifiés avec de la chitine colloïdale (Vu et al., 2009d).

1.4.6. Récupération des chitinases et de Zwittermicine A à partir de surnageant de fermentation de Bt pour produire des biopesticides avec une très forte activité biopesticide

Des SIW ont été complétés avec de la chitine colloïdale (0,2% p/v), et cette solution mixte (SIWC) a été utilisé comme milieu de fermentation pour produire des biopesticides à base de Btk HD-1 et ce, en se basant sur les résultats obtenus dans l'objectif 5. En outre, les SIW (sans supplément de chitine) ont été également employés comme matière première (comme du contrôle) pour la fermentation de Btk HD-1 pour comparer les résultats obtenus avec les SIWC. À la fin de la fermentation des SIW et SIWC, les pH des bouillons de la fermentation ont été ajustés à 4,5 : Ces bouillons étant centrifugés à 9000 g pendant 30 min pour obtenir des culots et des surnageants. Les culots des SIW et SIWC ont été mélangés avec leurs surnageants (respectifs) et le volume de chacun des suspensions des culots ainsi obtenu représentait un dixième du volume initial de suspension, c'est-à-dire que 10 L de bouillon de fermentation a donné 1L de suspension du culot (désignés comme Pel). La concentration en delta-endotoxines et spores, l'activité des chitinases et la Tx de la Pel ont été déterminées. La concentration des protéines solubles totales, les résidus de cellules et spores, l'activité des chitinases, et la présence de Zwittermicine A dans le surnageant de SIWC ont été évaluées.

Ultrafiltration du surnageant des SIWC: Le surnageant obtenu après centrifugation du bouillon de fermentation des SIWC a été ultra- filtré. Le matériel utilisé pour l'ultrafiltration est le Centramate ™ Tangentielle Flow Systems (numéro de catalogue OS005C12 - PALL Corporation). Le surnageant a été fractionné en deux parties par l'ultrafiltration: 1) la fraction de rétentat contenant des enzymes concentrées (chitinases, protéases alcalines), des spores et d'autres composantes de poids moléculaire plus élevé (comme les delta-endotoxines résiduelles; Vip3A); (2) la fraction du filtrat contenant la Zwittermicine A. Le prélèvement d'échantillons du surnageant, de retentate et du filtrat a été réalisé pour la mesure des paramètres (cellules et spores, activité des chitinases, protéines solubles totales). En outre, le surnageant et le filtrat des SIWC ont été utilisés pour identifier la présence de Zwittermicine A (méthode de Stabb et al., 1994; Silo-Suh et al., 1994)

Optimisation des ratios de mélange rétentat (Ret) et filtrat avec la suspension du culot des SIWC pour améliorer la Tx: Pour évaluer l'action synergique des chitinases et des autres composants (présents dans le rétentat) avec les complexes spores et delta-endotoxines, les rétentats ont été mélangés avec de la suspension du culot (Pel : Ret, par exemple, 1Pel-1Ret implique 1 volume de suspension du culot mélangé avec 1 volume de rétentat) dans différents ratios volumétriques, le tout a été soumis à des bioessais à l'aide de larves de la tordeuse des bourgeons de l'épinette pour estimer la valeur de Tx. L'effet synergique a été évalué sur la base de contrôle établi par le mélange de suspension du culot avec l'eau salée (Pel: Sal, volume: volume, 1Pel-1Sal implique 1 volume de suspension culot mélangés avec 1 volume d'eau salée) dans des ratios similaires. Afin d'évaluer l'action synergique de Zwittermicine A (présente dans le filtrat-Per) avec des complexes spores et delta-endotoxines, la suspension du culot a été mélangée au filtrat (Pel: Per, volume: volume) dans les différents ratios volumétriques et les mélanges ont été utilisés pour déterminer la Tx contre les larves de la tordeuse des bourgeons de l'épinette. L'effet synergique a été évalué sur la base de contrôle des mélanges de Sal: Pel, volume: volume. Les valeurs de la Tx de la suspension du culot des SIWC, de retentate, et du filtrat sont également comparées avec les résultats obtenus avec d'autres mélanges d'échantillons (Vu et al., 2009e).

1.4.7. Relations entre la concentration de delta-endotoxines et celle de spores, entre la concentration de spores et Tx ou entre

la concentration de delta-endotoxines et Tx de différents bouillons fermentés (milieu semi-synthétique, boues secondaire et SIW)

La fermentation de Btk HD-1 utilisant des boues secondaires comme matière première a été conduite en bioréacteur (15 L). Les paramètres tels que la concentration en cellules, en spores, en delta-endotoxines et les valeurs de Tx de bouillon de fermentation au cours de la fermentation ont été enregistrés. Ces résultats ont été comparés avec ceux obtenus à partir d'autres expériences utilisant les SIW et le milieu semi-synthétique (dans les objectifs de 2, 3, 5) pour analyser les relations entre la concentration en delta-endotoxines et celle en spores, entre la concentration en spores et la Tx ou entre la concentration en delta-endotoxines et la Tx de différents bouillons fermentés (Vu et al., 2009f).

1.5. RÉSULTATS ET DISCUSSION

Les résultats obtenus par ce projet sont présentés en 3 parties. La première décrit les résultats obtenus lors de l'utilisation de cinq méthodes d'amélioration de Tx. La seconde partie traite des résultats obtenus de la récupération par ultrafiltration de surnageants de bouillons de fermentation de chitinases et de la Zwittermicine A et de leur action synergique sur l'effet des complexes spores et delta-endotoxines pour la production de biopesticide avec une très forte activité insecticide. La troisième partie présente les relations entre la concentration de delta-endotoxines et celle de spores, entre la concentration de spores et la Tx, ou entre la concentration de delta-endotoxines et la Tx de différents bouillons fermentés.

1.5.1. Stratégies pour l'amélioration d'entomotoxicité (Tx) de biopesticide à base de *B. thuringiensis* en utilisant les eaux usées d'industrie d'amidon (SIW) comme matières premières

Cinq méthodes d'amélioration de Tx ont été utilisées pour la production de Btk HD-1 en utilisant les SIW comme substrat. Quatre méthodes visaient à fournir des éléments nutritifs dans les SIW avant la fermentation (suppléments de sources de carbone/azote appropriés dans les SIW) ou pendant la fermentation (enrichissement en nutriments via différents agents de contrôle du pH ou en utilisant le mode de la culture Fed-batch) ou l'augmentation des éléments nutritifs du substrat de fermentation (optimisation de la concentration des solides totaux de SIW). L'ajout d'éléments nutritifs dans les SIW, a augmenté les concentrations en cellules, en spores et en delta-endotoxines ainsi que les valeurs de Tx de bouillons fermentés et de Tx de suspensions des culots par rapport au contrôle (SIW sans modification). La cinquième méthode avait pour but de stimuler Btk HD-1 a produire plus de chitinases et a permis de démontrer que ces enzymes agissent comme agents de synergie pour accroître la valeur de Tx de bouillon fermenté ainsi que celle de la suspension du culot par rapport au contrôle (SIW sans chitine colloïdale). Le résumé des résultats obtenus de chaque méthode est inclus ci-dessous.

1.5.1.1. Impact de différents agents de contrôle du pH sur l'activité biopesticide de B. thuringiensis au cours de la fermentation en utilisant les SIW comme matières premières

Dans cette recherche, les différents agents de contrôle du pH ($NaOH/H_2SO_4$—SodSulp, $NaOH/CH_3COOH$—SodAcet, NH_4OH/CH_3COOH—AmmoAcet et NH_4OH/H_2SO_4—AmmoSulp) ont été utilisés pour maintenir le pH à 7 au cours de la fermentation de Btk HD-1 en utilisant les SIW comme matières premières. Il a été constaté qu'AmmoSulp et SodSulp sont les meilleurs agents de tampon du pH pour la production de protéases alcalines et d'amylases, respectivement, alors que le bouillon fermenté obtenu en utilisant SodAcet comme agents régulateurs avait la plus forte production de delta-endotoxines (1043,0 mg/L) et la plus haute valeur de Tx (18,4 x 10^9 SBU/l). La Tx de la suspension du culot (à un dixième du volume initial) en cas de SodAcet comme agents de contrôle du pH était de 26,7 x 10^9 SBU/l et la valeur plus élevée par rapport à trois autres agents de contrôle du pH. L'obtention de la plus grande concentration de delta-endotoxines et la valeur de Tx plus élevés en cas de SodAcet pourraient être dû à la présence de CH_3COOH pour contrôler le pH; CH_3COOH fournissant du carbone supplémentaire pour la croissance, la sporulation, la synthèse de delta-endotoxines, ce qui par la suite augmente la Tx. En outre, il a été constaté que NH_4OH a un impact négatif sur la croissance, la sporulation, la synthèse de delta-endotoxines et enfin la Tx de Btk HD-1, alors que l'ammoniaque a un effet positif sur la production de protéases alcalines. Ces travaux ont démontré que l'enrichissement de nutriments par des agents de contrôle du pH est une méthode simple et efficace pour accroître les valeurs de Tx du bouillon fermenté et de la suspension du culot (Vu et al., 2008).

1.5.1.2. Les eaux usées d'industrie d'amidon (SIW) pour la production de biopesticides - Effets des concentrations de matières solides.

Diverses concentrations de solides totaux (TS), allant de 15 à 66 g/L des eaux usées d'industrie d'amidon (SIW) ont été testées comme matières premières pour la production de biopesticide à base de Btk HD-1 par fermentation en erlenmeyers et en bioréacteurs de 15 L. La production en erlenmeyer a donné la plus forte concentration en delta-endotoxines à 30 g/L en TS et à 2,5% de volume de pré-culture (v/v). Les expériences en fermenteurs en utilisant les SIW (30 g/L en TS) ont été réalisées dans des conditions contrôlées de température, de pH et d'oxygène dissous et ont conduit à obtenir des concentrations plus élevées en spores, en delta-endotoxines et en enzymes (protéases et amylases), comparativement à ce qui est atteint avec les SIW (15 g/L en TS) comme contrôle. La valeur maximale de la Tx (bioessai contre larves de la tordeuse des bourgeons de l'épinette) des bouillons fermentés de Btk HD-1 a été observée avec des SIW 30 g/L en TS et suivie dans l'ordre: SIW 30 g/L TS (17,8 x 10^9 SBU/L) > SIW 15 g/L TS (SBU 15,3 x 10^9 SBU/ L) > milieu synthétique (11,7 x10^9 SBU/L).

De plus faibles concentrations en spores et en delta-endotoxines ont été observées avec les SIW 15g/L en TS que le milieu synthétique, mais les SIW 15 g/L en TS ont permis d'atteindre la plus grande valeur de la Tx. Dans ce cas, Btk HD-1 pourrait produire d'autres agents synergiques comme les protéines insecticides végétatives (Vips) et la Zwittermicine A à concentrations plus élevées au cours de la croissance dans les SIW bien qu'il pourrait synthétiser ces agents en concentrations plus faibles en milieu synthétique.

En outre, des suspensions des culots (un dixième du volume initial) de trois milieux ont été préparées et ont été utilisés pour déterminer les valeurs de Tx. Les valeurs de Tx de suspensions des culots à partir de trois milieux s'établissaient dans l'ordre suivant: SIW 30 g/L en TS (25,8 x 10^9 SBU/L)> milieu synthétique (22,7 x10^9 SBU/L)> SIW 15 g/L en TS (20,9 x 10^9 SBU/L). Ces recherches ont démontré que les SIW 15 g/L en TS (contrôle) n'ont pas suffisamment de nutriments pour maximiser la production de biopesticide avec une très fort Tx. Par conséquent, les SIW 30 g/L en TS peuvent être utilisés dans l'avenir comme substrats de fermentation pour la production de biopesticides à base de Btk HD-1 (Vu et al., 2009a).

1.5.1.3. Production de biopesticide à base de B. thuringiensis en utilisant les SIW enrichis de différentes sources de carbone/azote comme milieux de fermentation

Différentes sources de carbone et d'azote ont été ajoutées dans les SIW à diverses concentrations et ces mélanges ont été utilisés comme milieux pour la production de Btk HD-1 par fermentation en erlenmeyers. Les concentrations en cellules, en spores et en delta-endotoxines (dans des bouillons fermentés de 48h) ont été évaluées pour déterminer quels types de sources de carbone/azote et à quelles concentrations doivent-ils être ajoutés dans les SIW pour obtenir des concentrations élevées de ces paramètres. L'amidon de maïs (comme source de carbone), ajouté aux SIW à une concentration finale de 1,25%, p/v a conduit à obtenir des concentrations plus élevées en spores et en delta-endotoxines par rapport aux SIW contenant d'autres concentrations d'amidon de maïs ou d'autres sources de carbone. Les SIW fortifiées avec du Tween 80 à 0,3 ou 0,4%, v/v ont donné la plus grande concentration en delta-endotoxines par rapport aux SIW complétées avec d'autres concentrations de Tween 80. Dans le cas d'apport d'azote, les SIW complétées avec différentes sources n'ont pas permis d'augmenter significativement les concentrations en cellules, en spores et en delta-endotoxines, l'ajout d'azote a même entraîné des résultats négatifs en concentrations plus élevées aux SIW. En outre, les SIW complétées avec deux sources de carbone soit l'amidon de maïs à 1,25%, p/v et le Tween 80 à 0,2 ou 0,3%, v/v ont permis d'atteindre des concentrations plus élevées en cellules, en spores et en delta-endotoxines par rapport à ces paramètres obtenus dans les SIW complétées avec l'amidon de maïs à 1,25% (p / v) et le Tween 80 à 0,1 ou 0,4%, v/v.

La fermentation de Btk HD-1 en utilisant les SIW complétées avec l'amidon de maïs à 1,25%, p/v et le Tween 80 à 0,2%, v/v (désignée comme SIWST) a été effectuée en bioréacteur de 15 L dans des conditions contrôlées (pH, oxygène dissous, etc.) La concentration en delta-endotoxines, les valeurs de la Tx du bouillon fermenté et de la Tx d'une suspension du culot des SIWST atteignaient 1327 µg/ml; 18,1 x 10^6 SBU/ml et 26,7 x 10^6 SBU/ml, respectivement. Ces paramètres sont significativement plus élevés que ce qui est atteint par l'utilisation des SIW- contrôle. Ces recherches ont confirmé que l'amidon de maïs et le Tween 80 peuvent non seulement servir de source de carbone mais aussi comme agents pouvant améliorer le transfert de l'oxygène au cours de la fermentation, ceci augmentant de manière significative les concentrations en cellules, en spores et en delta -endotoxines ainsi que la Tx du bouillon fermenté et celle d'une suspension du culot (Vu et al., 2009b).

1.5.1.4. Fermentation en Fed-batch de B. thuringiensis en utilisant les SIW comme milieux de fermentation

Des biopesticides à base de Btk HD-1 ont été produits par fermentation en batch et en Fed-batch en employant des SIW comme seuls substrats. Le mode de culture Fed-batch a été conduit en se basant sur l'oxygène dissous (DO) comme paramètre de contrôle rétroactif et le mode manuel d'intermittence de l'alimentation en éléments nutritifs présents dans les SIW dans le milieu de fermentation. Il a été démontré que la fermentation en Fed-batch avec deux alimentations intermittentes (à 10 et 20 h) au cours d'une fermentation de 72 heures a donné le maximum de la concentration en delta-endotoxines (1672,6 μg/ml) et de la valeur de la Tx (18,5 x 10^6 SBU/ml) dans le bouillon fermenté et donc significativement plus élevées que les maximum en delta-endotoxines (511,0 μg/ml) et Tx (15,8 x 10^6 SBU/ml) obtenus par fermentation en batch ou d'autres fermentations en Fed-batch (avec une ou trois alimentations).

La fermentation en Fed-batch avec trois alimentations intermittentes (à 10, 20 h et 34 h) pourrait améliorer de façon significative la concentration en cellules, mais il y a eu apparition de Btk non sporogènes qui a provoqué une baisse de la production de spores, de delta-endotoxines et de la valeur de la Tx du bouillon fermenté.

Les valeurs de Tx de suspensions des culots des fermentations en batch et en Fed-batch variaient selon l'ordre suivant: fermentation en Fed-batch avec deux alimentations (27.4 x 10^6 SBU/ml) > fermentation en Fed-batch avec un apport (22.7 x 10^6 SBU/ml) ≥ fermentation en Fed-batch avec trois ajouts (21.8 x 10^6 SBU/ml) ≥ fermentation en batch (21.0 x 10^6 SBU/ml) (pour Fisher ($P < 0.05$, Statistica 7.0)). La recherche a confirmé que (1) le profil en ligne de l'oxygène dissous lors de la culture de Btk HD-1 peut servir d'indicateur permettant d'identifier les périodes pour alimentations et (2) la fermentation en Fed-batch avec deux ajouts intermittents (à 10 h et 20 h) paraissant la meilleure stratégie pour améliorer la production de cellules, de spores et de delta-endotoxines et, par conséquent, augmenter les valeurs des Tx du bouillon fermenté et de la suspension du culot.

En outre, il a été constaté que des élévations en concentrations en spores ou en delta-endotoxines n'augmentent pas proportionnellement la Tx du bouillon fermenté. Les valeurs de la Tx spécifique (SBU/µg delta-endotoxine, SBU/1000 spores) varient avec le mode de batch ou le mode de Fed-batch ou le nombre d'alimentations dans le mode de Fed-batch (Vu et al., 2009c).

1.5.1.5. Production induite des chitinases pour améliorer l'entomotoxicité de B. thuringiensis en utilisant les SIW comme substrat de fermentation

La production induite des chitinases au cours de la bioconversion des SIW en biopesticides à base de Btk HD-1 a été étudiée par fermentation en erlenmeyers et en bioréacteurs. Des SIW ont été fortifiées avec des concentrations différentes (0%, 0,05%, 0,1%, 0,2%, 0,3% p/v) de chitine colloïdale et l'effet de ces additions a été constaté en termes de concentrations en cellules et en spores et des activités des chitinases, des protéases et des amylases et de la valeur de la Tx. À la concentration optimale de chitine colloïdale (0,2% p/v), la valeur de la Tx du bouillon fermenté et celle de suspension du culot ont été améliorées de 12,4 x 10^9 (sans la chitine) à 14,4 x 10^9 SBU/L et de 18,2 x 10^9 (sans la chitine) à 25,1 x 10^9 SBU/L, respectivement.

En outre, des expériences ont été effectuées en bioréacteurs de 15 L avec des SIW enrichies de chitine colloïdale à 0,2% p/v (désigné comme SIWC) pour induire la production de chitinases par Btk HD-1 et comparativement l'emploi des SIW sans chitine colloïdale comme contrôle. Il a été constaté que les concentrations en cellules, en spores et en delta-endotoxines et les activités des protéases et des amylases ont été réduites alors que l'entomotoxité et l'activité des chitinases ont été augmentées dans le bouillon fermenté des SIWC; l'activité des chitinases atteignant une valeur maximale à 24 h (15 mU/ml) et la Tx de suspension du culot atteint le niveau le plus élevé (26,7 x 10^9 SBU/L) à 36 h de fermentation dans les SIWC. Dans le contrôle (SIW sans chitine colloïdale), la valeur maximale de Tx de la suspension du culot (20,5 x 10^9 SBU / L) est atteint après 48 h de fermentation. Un effet synergique des delta-endotoxines, des spores et de l'activité de chitinases sur la Tx contre les larves de la tordeuse des bourgeons de l'épinette a été observé.

Il a été constaté par ailleurs que des Tx à l'égard de spores (SpTx-spore, SBU/1000 spores) et des delta-endotoxines (SpTx toxine, SBU/µg delta-endotoxines) ont été significativement plus élevées dans les SIWC que dans les SIW. Ce fait démontre qu'il existe une synergie d'action de spores, dse delta-endotoxines et de l'activité de chitinases exprimée en termes de valeur de la Tx (Vu et al., 2009d).

1.5.1.6. Comparaison des résultats obtenus à partir de cinq méthodes

Par combinaison des résultats obtenus dans les cinq méthodes (Tableau 5), quelques remarques s'imposent:

- L'augmentation de la concentration en spores accroît la concentration en delta-endotoxines, sauf en cas d'utilisation NaOH/CH_3COOH comme agents de contrôle du pH, la concentration de spores a été aussi faible que dans les SIW (contrôle), mais celle en delta-endotoxines est sensiblement plus élevée que le contrôle. Ceci pourrait être due au rôle de CH_3COOH comme une source de carbone favorable à la synthèse de la delta-endotoxine de Btk HD-1 (Zouari et Jaoua, 1999) ou bien ce bacille pourrait produire plus d'un cristal protéique insecticide (ICPs) par cellule sporulée, lorsqu'il est cultivé dans les SIW en présence de CH_3COOH, ceci causant l'augmentation de la concentration de delta-endotoxines. En fait, il est connu que (1) Btk HD-1 peut produire deux inclusions: une bipyramidale, l'autre ovoïde (Aronson et al., 1986); (2) certaines cellules peuvent ne pas inclure l'inclusion ovoïde et plus d'un cristal bipyramidal peuvent être synthétisés au sein d'une cellule. Par contre la synthèse d'une seule inclusion ovoïde par sporange n'a jamais été observée (Bechtel and Bulla, 1976).

- Les valeurs maximales des Tx obtenues dans les cinq bouillons fermentés (où cinq méthodes ont été appliquées) ne sont pas significativement différentes les unes des autres, mais elles étaient sensiblement plus élevées que celle obtenue en bouillon fermenté des SIW-contrôles. De même, les valeurs maximales des Tx obtenues dans les cinq suspensions des culots (où cinq méthodes ont été appliquées) ne sont pas significativement différentes les unes des autres, mais elles étaient également sensiblement plus élevées que la Tx obtenue en suspension de pellet des SIW (de contrôle) (tableau 5).

- Les valeurs maximales des Tx obtenues dans les cinq suspensions des culots (où cinq méthodes ont été appliquées) ne sont pas significativement différentes les unes des autres même si les concentrations en spores et en delta-endotoxines dans ces suspensions des culots sont divergent. Par conséquent, il est impossible de confirmer que les biopesticides-Bt qui contiennent des concentrations élevées en spores ou en delta-endotoxines auront des Tx sensiblement plus élevées que les autres produits à faible titrage en spores et delta-endotoxines. Les raisons possibles sont : (1) la présence d'agents solubles tels que les protéines insecticides végétatives -Vips, Zwittermicin A, etc dans ces suspensions des culots (à différentes concentrations), qui est finalement affecté à leur valeurs de Tx; (2) les delta-endotoxines ou les spores produites par Btk HD-1 dans différents milieux n'auront pas le même potentiel toxique pour la même cible d'insectes ravageurs (Liu et Bajpai, 1995). Cette hypothèse est clairement démontrée en terme de SpTx-spores (Tx spécifiques par

1000 spores) ou en terme de SpTx-toxine (Tx spécifiques par microgrammes de delta-endotoxines) (Tableau 5). Les valeurs de SpTx-spore ou SpTx-toxines sont significativement différentes entre les bouillons fermentés ou entre les suspensions des culots obtenus dans les divers milieux (modifié) pour la fermentation. Les valeurs les plus élevées de SpTx-spores et de SpTx-toxine apparaissent dans SIWC, cette réalité en outre confirme le rôle très important des chitinases dans le renforcement de la Tx (même à faible concentration de spores et de delta-endotoxines par rapport à d'autres médias).

- Il a également été constaté que les valeurs maximales des Tx du bouillon fermenté ou des Tx de suspension du culot de SIWC surviennent à 36 h de fermentation, tandis que dans d'autres cas, les valeurs maximales des Tx sont obtenues après 48 h (tableau 5); La rapidité de la fermentation étant un des paramètres avantageux pour la production économique de biopesticides.
- En outre, il convient de mentionner que les chitinases et autres composants (delta-endotoxines résiduelles, spores, Vips, etc.) sont présents dans le surnageant après centrifugation du bouillon fermenté, surnageant qui est généralement jeté. Par conséquent, les chitinases et les autres composants seront perdus. Cette réalité conduit à l'idée de la nécessité de récupérer par ultrafiltration tous ces agents de synergie contenus dans le surnageant. Les résultats de la récupération des chitinases et autres composants sont présentés ci-dessous.

1.5.2. Action des chitinases et de la Zwittermicine A comme agents de synergie de l'effet des delta-endotoxines et des spores de *B. thuringiensis* contre la tordeuse des bourgeons de l'épinette

- Cette partie présente la récupération des chitinases et de la Zwittermicine A à partir de surnageant du bouillon fermenté des SIWC par des processus d'ultrafiltration. Il a été montré que les chitinases et la Zwittermicine A récupérés ont une action synergique pour l'activité des delta-endotoxines et des spores contre les larves de la tordeuse des bourgeons de l'épinette (*Choristoreuna fumiferena*) ce qui par conséquent, accroît de manière efficace la récupération de Tx de bouillon fermenté.

1.5.2.1. Récupération des chitinases et de Zwittermicine A à partir de surnageant de fermentation de B. thuringiensis pour produire des biopesticides ayant une très forte activité insecticide

Le surnageant du bouillon fermenté de SIWC contenant des chitinases et de la Zwittermicine A. a été ultra - filtré sur membrane ayant un seuil de coupure de 5kDa coupure, fonctionnant à un flux de 480 L/h/m^2 et sous une pression transmembranaire de 75 kPa. Les récupérations de protéines solubles, de chitinases et de spores dans le rétentat ont atteint respectivement 79,4%, 50% et 56,7%. La Zwittermicine A a traversé la membrane et a été recueillie dans le filtrat. Le rétentat (contenant les chitinases et autres composants concentrés) et le filtrat ont été mélangés séparément avec des suspensions du culot de SIWC selon divers ratios volumétriques pour obtenir différentes concentrations en delta-endotoxines, en spores et en activité chitinolytique et ces mélanges ont été soumis à des bioessais contre la tordeuse des bourgeons de l'épinette les larves (*Choristoreuna fumiferena*). Il a été démontré que les chitinases (et autres composants) ou de la Zwittermicine A ont un effet synergique sur les delta-endotoxine, les spores et la Tx contre les larves de la tordeuse des bourgeons de l'épinette.

Ces travaux ont révélé que le mélange de la suspension du culot et du rétentat dans un ratio volumétrique de 1:4 (1Pel4Ret) permet la plus haute récupération de Tx (65,1%) avec le plus gros volume de 5L (1L suspension du culot et 4L retentate). Ce mélange contient des concentrations respectives

en spores, en delta-endotoxines et en activité chitinolytique de 3,24 x 10^8 (UFC/ml), 1,09 (mg/ml) et 69,3 (mU/ml). La Tx de ce mélange a été de 22,8 x 10^6 (SBU/ml), qui a été le plus efficace ratio de mélange volumétrique en terme d'action synergique des chitinases (et autres éléments possibles) sur les delta-endotoxines et les spores ainsi que sur la Tx.
En outre, le mélange de suspension du culot avec du filtrat dans un ratio volumétrique de 1:2 (1Pel2Per) a donné la plus forte action synergique de la Zwittermicine A dans le filtrat avec les delta-endotoxines et les spores sur Tx. Ce mélange contient des concentrations de spores, de delta-endotoxines, d'activité des chitinases et de la Tx de 5,40 x 10^8 (CFU/ml), 1,82 (mg/ml), 10,2 (mU/ml), et de 22,8 x 10^6 (SBU/ ml), respectivement. Cependant, ce mélange est moins efficace en termes de récupération de la Tx du bouillon fermenté des SIWC par rapport au mélange d'une partie de suspension du culot pour 4 parties de rétentat (Vu et al., 2009e).

1.5.3. Relations de divers paramètres (spores, delta-endotoxines et Tx)

1.5.3.1. Relations entre la concentration de delta-endotoxines et celle de spores, entre la concentration de spores et la Tx ou entre la concentration de delta-endotoxines et la Tx de différents bouillons fermentés

Six différents milieux semi-synthétiques, les SIW et les boues secondaires ont été utilisées pour la production par fermentation de Btk HD-1. Les concentrations de spores, de delta-endotoxines et l'activité biopesticide (entomotoxicité-Tx) obtenues dans ces expériences ont été enregistrées et les relations possibles entre ces paramètres analysées. Il est intéressant de constater que la relation entre la concentration de delta-endotoxines et celle des spores dans les différents milieux suit l'équation suivante: **concentration de delta-endotoxines = a (concentrations de spores)^b, avec b > 0, $R^2 \geq$ 0.90**. Cette relation dépend de chaque milieu (constantes "a et b sont différents). La relation entre SpTX-spore (Tx spécifiques par 1000 spores) et la concentration de spores suit strictement l'équation: **SpTx-spore = c (concentrations de spores) ^{d}, avec d < 0, R^2 > 0.98**. Cette relation conduit à la remarque que (1) des spores produites au début de la période de fermentation peuvent être plus toxiques que celles produites en fin de fermentation, indépendamment des milieux ou (2) en d'autres termes, Btk HD-1 pourrait produire des agents de synergie (Vip3A, Zwittermicin A, etc), à plus forte concentration au début de la période de fermentation; ces agents contribuant à la valeur Tx.

La relation entre Tx et la concentration de delta-endotoxines suit une relation linéaire: **Ln (Tx) = m (concentration de delta-endotoxin) + h**; 'm' étant la pente et 'h' l'ordonnée à l'origine. Un examen plus approfondi des résultats en termes de valeurs de Tx, de spores, de concentration en delta-endotoxines des suspensions des culots (complexe de spores et de delta-endotoxines) de ces milieux, il a été constaté que la différence entre les variations des valeurs de Tx par rapport aux diverses concentration en spores ou en delta-toxine sont dus seulement à des facteurs solubles; Ces facteurs étant présents dans les surnageants de bouillons fermentés centrifugés (Vu et al., 2009f).

1.6. REFERENCES

Ackermann H.W., Azizbekyan R.R., Bernier R.L. , de Barjac H., Saindoux S., Valdro J.R, and Yu M.X. (1995) Phage typing of *Bacillus subtilis* and *B. thuringiensis*. Res. Microbiol., 146: 643-657.

Agaisse H. and Lereclus D. (1995) How Does *Bacillus thuringiensis* produce so much insecticidal crystal protein? J. Bacteriol., 177: 6027-6032.

Agaisse H., Gominet M., Okstad O.A., Kolsto A.B., and Lereclus D. (1999) PlcR is a pleiotropic regulator of extracellular virulence factor gene expression in *Bacillus thuringiensis*. Mol. Microbiol., 32: 1043-1053.

Andrews R.E.Jr., Bibilos M.M., and Bulla L.A.Jr. (1985) Protease activation of the entomocidal protoxin of *Bacillus thuringiensis* subsp. *kurstaki*. Appl. Environ. Microbiol., 50: 737-742.

Tableau 5.Comparaison des résultats obtenus à partir de cinq méthodes appliquées pour améliorer la Tx

Paramètres	**Contrôle**	**Méthodes appliquées pour améliorer la Tx**				
	SIW (15 g/L TS)	Agents de contrôle du pH ($NaOH/CH_3COOH$)	Concentration maximale des solides totaux [**SIW 30 g/L TS**]	Fortification par différentes sources de carbone/azote dans les SIW (SIW + amidon de maïs (1.25%, w/v) + Tween 80 (0.2%, v/v)) [**SIWST**]	Fermentation en Fed-batch (avec deux alimentations)	Chitinases comme agent de synergie (SIW + chitine colloïdale (0.2 %, p/v)) [**SIWC**]
Concentration maximale des spores dans le bouillon fermenté (x 10^8 CFU/ml) [A]	1.50 ± 0.10 [d] (24 h)	1.43 ± 0.09 [d] (24 h)	3.0 ± 0.21 [c] (36 h)	6.0 ± 0.4 [b] (48 h)	13.00 ± 0.08 [a] (48 h)	1.20 ± 0.07 [e] (24 h)

Concentration maximale des Delta-endotoxines dans le bouillon (µg/ml) [A]	459.5 ± 22.9 [e] (48 h)	1043.0 ± 47.8 [d] (48 h)	1112.1 ± 55.6 [c] (48 h)	1327.0 ± 66.4 [b] (48 h)	1672.6 ± 83.9 [a] (48 h)	407.55 ± 19.15 [ef] (48 h)
Tx maximale dans bouillon fermenté (x 10^6 SBU/ml) [A]	15.3 ± 1.0 [b] (48 h)	18.40 ± 1.38 [a] (48 h)	17.80 ± 1.20 [a] (48 h)	18.10 ± 1.20 [a] (48 h)	18.50 ± 1.50 [a] (48 h)	17.1 ± 1.15 [ab] (36 h)
SpTx-spore (SBU/1000 spores) du bouillon fermenté	102	153.3	59.3	30.2	14.2	240.8
SpTx-toxine (x 10^3 SBU/µg delta-endotoxine) du bouillon fermenté	33.3	17.64	15.8	13.6	11.0	42.6
Concentration maximale des spores dans la suspension du culot (x10^8 CFU/ml) [A]	14.2 ± 0.9	11.4 ± 0.9	28.5 ± 1.7	57.0 ± 4.0 [a]	124.0 ± 8.6 (48h)	6.45 ± 0.42
Concentration maximale des delta-endotoxines	4595.0 ± 229	10430 ± 478	11121 ± 556	13270 ± 664 [a]	16726 ± 839 (48 h)	4019.4 ± 172.8

dans la suspension du culot (µg/ml) [B]						
Tx maximale de suspension du culot (x 10^6 SBU/ml) [A]	20.9 ± 1.5 [b] (48 h)	26.70 ± 2.10 [a] (48 h)	25.80 ± 1.80 [a] (48 h)	26.70 ± 1.90 [a] (48 h)	27.4 ± 1.90 [a] (48 h)	26.7 ± 2.1[a] (36 h)
SpTx-sporee (SBU/1000 spores) de suspension du culot	14.7	23.4	9.1	4.7	2.2	39.6
SpTx-toxine (x 10^3 SBU/µg delta-endotoxines) de suspension du culot	4.5	2.56	2.3	2.0	1.6	6.64

[A] Chaque résultat est la moyenne des trois répétitions. Dans chaque ligne, la lettre qui est près du nombre, est l'indicateur le moins significatif au point de vue différence ($P < 0.05$, Statistica 7.0).

[B] Concentration des delta-endotoxines en suspension du culot provenant de celle des delta-endotoxines dans le bouillon fermenté qui a été concentré 10 fois.

Annachhatre A.P., and Amatya P.L. (2000) UASB treatment of tapioca starch wastewater. J. Environ. Eng., 126: 1149-1152.

Arcas J., Yantorno O., and Ertola R. (1987) Effect of high concentration of nutrients on *Bacillus thuringiensis* cultures. Biotechnol. Lett., 9: 105-110.

Aronson A.I., Beckman W., and Dunn P. (1986) *Bacillus thuringiensis* and related insect pathogens. Microbiol. Rev., 50: 1-24.

Aronson A.I., and Shai Y. (2001) Why *Bacillus thuringiensis* insecticidal toxins are so effective: unique features of their mode of action. FEMS Microbiol. Lett., 195: 1-8.

Aronson A.I. (2002) Sporulation and δ-endotoxin synthesis by *Bacillus thuringiensis*. Cell. Mol. Life Sci., 59: 417-425.

Arora N., Ahmad T., Rajagopal R. and Bhatnagar R.K. (2003) A constitutively expressed 36 kDa exochitinase from *Bacillus thuringiensis* HD-1. Biochem. Biophys.Res. Com., 307: 620–625.

Barboza-Corona J.E., Contreras J.C., Velázquez-Robledo R., Bautista-Justo M., Gómez-Ramírez M., Cruz-Camarillo R. and Ibarra J.E. (1999) Selection of chitinolytic strains of *Bacillus thuringiensis*. Biotechnol.Lett., 21: 1125–1129.

Barboza-Corona J.E., Nieto-Mazzocco E., Velázquez-Robledo R., Salcedo-Hernández R., Bautista M., Jiménez B. and Ibarra J.E. (2003) Cloning, sequencing, and expresion of the chitinase gene chiA74 from *Bacillus thuringiensis*. Appl. Environ. Microbiol., 69: 1023-1029.

Barnabé S. (2004) Bioconversion des boues d'épuration en biopesticides: utiliser le plein potentiel nutritif du substrat pour une meilleure valeur ajoutée (entomotoxicité). Ph.D. thesis, INRS-ETE, Université du Québec, Québec, Canada.

Bechtel D.B., and Bulla L.A.Jr. (1976) Electron Microscope Study of Sporulation and Parasporal Crystal Formation in *Bacillus thuringiensis*. J. Bacteriol., 127: 1472-1481.

Ben-Dov E., Zaritsky A., Dahan E., Barak Z., Sinai R., Manasherob R., Khamraev A., Troitskaya E., Dubitsky A., Berezina N., and Margalith Y. (1997). Extended screening by PCR for seven *cry*-group genes from field-collected strains of *Bacillus thuringiensis*. Appl. Environ. Microbiol., 63: 4883-4890.

Bernhard K., Jarrett P., Meadows M., Butt J., Ellis D.J., Roberts G.M., Pauli S., Rodgers P., and Burges H.D. (1997) Natural isolates of *Bacillus thuringiensis*: worldwide distribution, characterization, and activity against insect pests. J. Invertebr. Pathol., 70: 59-68.

Brar S.K., Verma M., Tyagi R.D., Valéro J.R., and Surampalli R.Y. (2005a) Impact of Tween 80 during *Bacillus thuringiensis* fermentation of wastewater sludges. Process Biochem., 40: 2695-2705.

Brar S.K., Verma M., Tyagi R.D., Valéro J.R., and Surampalli R.Y. (2005b) Starch Industry Wastewater-Based Stable *Bacillus thuringiensis* Liquid Formulations. J. Econ. Entomol. 98: 1890-1898.

Brar S.K., Verma M., Tyagi R.D., Valéro J.R., and Surampalli R.Y. (2006a) Efficient centrifugal recovery of Bacillus thuringiensis biopesticides from fermented wastewater and wastewater sludge. Water Res., 40: 1310 - 1320.

Brar S.K., Verma M., Tyagi R.D., Valéro J.R., and Surampalli R.Y. (2006b) Screening of Different Adjuvants for Wastewater/Wastewater Sludge-Based *Bacillus thuringiensis* Formulations. J. Econ. Entomol., 99: 1065-1079.

Brar S.K., Verma M., Tyagi R.D., Surampalli R.Y., Barnabé S., and Valéro J.R. (2007) *Bacillus thuringiensis* proteases: Production and role in growth, sporulation and synergism. Process Biochem., 42: 773-790.

Brar S.K. (2007) Effets des propriétes rhéologiques sur la fermentation des eaux usées et des boues d'épuration par Bacillus thuringiensis var. kurstaki et sur le développement de biopesticides en suspensison aqueouses concentrées. Ph.D. thesis, INRS-ETE, Université du Québec, Québec, Canada.

Bravo A., Sarabia S., Lopez L., Ontiveros H., Abarca C., Ortiz A., Ortiz M., Lina L., Villalobos F.J., Peña G., Nuñez-Valdez M.-E., Soberon M., and Quintero R. (1998) Characterization of *cry* genes in a mexican *Bacillus thuringiensis* strain collection. Appl. Environ. Microbiol. 64: 4965-4972.

Bravo A., Gill S.S., and Soberón M. (2007) Mode of action of Bacillus thuringiensis Cry and Cyt toxins and their potential for insect control. Toxicon, 49: 423-435.

Broderick N.A., Goodman R.M., Raffa K.F. and Handelsman J. (2000) Synergy Between Zwittermicin A and *Bacillus thuringiensis* subsp. *kurstaki* Against Gypsy Moth (Lepidoptera: Lymantriidae). Environ. Entomol. 29: 101-107.

Broderick N.A., Goodman R.M., Handelsman J., Raffa K.F. (2003) Effect of host diet and insect source on synergy of gypsy moth (Lepidoptera: Lymantriidae) mortality to *Bacillus thuringiensis* subsp. *kurstaki* by zwittermicin A. Environ. Entomol, 32: 387-391.

Bulla L.A.Jr., Rhodes R.A., and Julian G.St. (1975) Bacteria as insect pathogen. Ann. Rev. Microbiol., 29: 163-190.

Bulla L.A.Jr., Kramer K.J., and Davidson L.I. (1977) Characterization of the entomocidal parasporal crystal of *Bacillus thuringiensis*. J. Bacteriol., 130: 375-383.

Bulla L.A.Jr., Kramer K.J., Cox D.J., Jones B.L., Davidson L.I., and Lookhart G.L. (1981) Purification and characterization of the entomocidal protoxin of *Bacillus thuringiensis*. J. Biol. Chem., 256: 3000-3004.

Campbell D.P., Dieball D.E. and Brackett J.M. (1987) Rapid HPLC Assay for the β-Exotoxin of *Bacillus thuringiensis*. J. Agric. Food chem., 35: 156-158.

Carlson C.R., and Kolsto A.B. (1993) A Complete Physical Map of a *Bacillus thuringiensis* Chromosome. J. Bacteriol., 175: 1053-1060.

Carlson C.R., Caugant D.A., and Kolsto A.B. (1994) Genotypic Diversity among *Bacillus cereus* and *Bacillus thuringiensis* Strains. Appl. Environ. Microbiol., 60: 1719-1725.

Carozzi N.B., Kramer V.C., Warren G.W., Evola S., and Kozil M.G. (1991) Prediction of insecticidal activity of *Bacillus thuringiensis* strains by polymerase chain reaction product profiles. Appl. Environ. Microbiol., 57: 3057-3061.

Chen S., Hong J.Y., and Wu W.T. (2003) Fed batch culture of *Bacillus thuringiensis* based on motile intensity. Journal of Industrial Microbiology and Biotechnology, 30: 677-681.

Chestukhina, G. G., I. A. Zalunin, L. I. Kostina, T. S. Kotova, S. P. Katrukha, and V. M. Stepanov. 1980. Crystal-forming proteins of *Bacillus thuringiensis*: the limited hydrolysis by endogenous proteinases as a cause of their apparent multiplicity. Biochem. J. 187:457–465.

Colin X., Farinet J.-L., Rojas O., and Alazard D. (2007) Anaerobic treatment of cassava starch extraction wastewater using a horizontal flow filter with bamboo as support. Bioresource Technology 98: 1602-1607.

Crickmore N., Zeigler D.R., Feitelson J., Schnepf E., Van Rie J., Lereclus D., Baum J., and Dean D.H. 1998. Revision of the nomenclature for the *Bacillus thuringiensis* pesticidal crystal proteins. Microbiol. Mol. Biol. Rev., 62: 807-813.

Cox G.M., McDade H.C., Chen S.C.A, Tucker T.C., Gottfredsson M., Wright L.C., Sorrell T.C., Leidich S.D., Casadevall A., Ghannoum M.A., and Perfect J.R. (2001) Extracellular phospholipase activity is a virulence factor for *Cryptococcus neoformans*. Molecular Microbiology, 39(1) 166-175.

Dalhammar G., and Steiner H. (1984) Characterization of inhibitor A, a protease from *Bacillus thuringiensis* which degrades attacins and cecropins, two classes of antibacterial proteins in insects. Eur. J. Biochem., 139: 247-252.

de Barjac H., and Bonnefoi A. (1973) Mise au point sur la classification des Bacillus thuringiensis. Entomophaga, 18: 5-17.

de Barjac H., and Frachon E. (1990) Classification of *Bacillus thuringiensis* strains. Entomophaga, 35: 233-240.

de Maagd R.A., Bravo A., and Crickmore N. (2001) How Bacillus thuringiensis has evolved specific toxins to colonize the insect world. Trends Genet., 17: 193-199.

Donovan W.P., Tan Y., and Slaney (1997) Cloning of the *nprA* Gene for Neutral Protease A of *Bacillus thuringiensis* and Effect of In Vivo Deletion of *nprA* on Insecticidal Crystal Protein. Applied and Environmental Microbiology 63: 2311-2317.

Donovan, W. P., Donovan, J. C., and Engleman, J. T. (2001) Gene knockout demonstrates hat *vip3A* contributes to the pathogenesis of *Bacillus thuringiensis* toward *Agrotis ipsilon* and *Spodoptera frugiperda. J. Invertebr. Pathol.*, 78, 45–51.

Driss F., Kallassy-Awad M., Zouari N. and Jaoua S. (2005) Molecular characterization of a novel chitinase from *Bacillus thuringiensis* subsp. *kurstaki*. Journal of Applied Microbiology 99: 945–953.

Dubois N.R., and Dean D.H. (1995) Synergism between Cry1A insecticidal crystal proteins and spores of *Bacillus thuringiensis*, other bacterial spores and vegetative cells against *Lymantria dispar* (Lepidoptera: Lymantriidae) larvae. *Environ Entomol* 24, 1741-1747.

Dulmage H. (1970) Insecticidal activity of HD-1, a new isolate of Bacillus thuringiensis var. alesti. Journal of Invertebrate Pathology 15: 232-239.

Dulmage H.T., Correa J.A., and Martinez A.J. (1970) Coprecipitation with Lactose as a Means of Recovering the Spore-Crystal Complex of BaciZZus thuringiensis. Journal of Invertebrate Pathology 15: 15-20.

Edlund T., Siden I., and Boman H.G. (1976) Evidence for two immune inhibitors from *Bacillus thuringiensis* interfering with the humoral defense system of saturniid pupae. Infect. Immun., 14: 934-941.

Espinasse S., Chaufaux J., Buisson C., Perchat S., Gohar M., Bourguet D., Sanchis V. (2003) Occurrence and Linkage Between Secreted Insecticidal Toxins in Natural Isolates of *Bacillus thuringiensis*. Current Microbiology 47: 501-507.

Estruch, J. J., Warren, G. W., Mullins, M. A., Nye, G. J., Craig, J. A., and Koziel, M. G. (1996) Vip3A, a novel *Bacillus thuringiensis* vegetative insecticidal protein with a wide spectrum of activities against lepidopteran insects. *Proc.Natl. Acad. Sci. USA*, 93, 5389–5394.

Farrera F., Perez-Guevara F., and de la Torre M. (1998) Carbon: nitrogen ratio interacts with initial concentration of total solids on insecticidal crystal protein and spore production in *Bacillus thuringiensis* HD-73. Applied Microbiology and Biotechnology, 49: 758-765.

Federici B.A. (2005) Insecticidal bacteria: An overwhelming success for invertebrate pathology. Journal of Invertebrate Pathology, 89: 30-38.

Fedhila S., Nel P., and Lereclus D. (2002) The InhA2 Metalloprotease of *Bacillus thuringiensis* Strain 407 Is Required for Pathogenicity in Insects Infected via the Oral Route. Journal of Bacteriology, 184(12): 3296-3304.

Gaviria Rivera A.M. and Fergus G. Priest (2003) Molecular Typing of *Bacillus thuringiensis* Serovars by RAPD-PCR. System. Appl. Microbiol., 26 : 254-261.

Gerhardt P., Pankratz H.S. and Scherrer R. (1976) Fine structure of the bacillus thuringiensis spore. Applied and Environmental Microbiology, 32(3): 438-440.

Gilmore M.S., Cruz-Rodz A.L., Leimeister-Wachter M., Kreft J., and Goebel W. (1989) A Bacillus cereus Cytolytic Determinant, Cereolysin AB, Which Comprises the Phospholipase C and Sphingomyelinase

Genes: Nucleotide Sequence and Genetic Linkage. Journal of Bacteriology 171(2):744-753.

Gohar M., Gilois N., Graveline R., Garreau C., Sanchis V., and Lereclus D. (2005) A comparative study of Bacillus cereus, Bacillus thuringiensis and Bacillus anthracis extracellular proteomes. Proteomics, 5: 3696-3711.

Goldberg I., Sneh B., Battat E., and Klein D. (1980) Optimisation of a medium for a high yield production of spore-crystal preparation of Bacillus thuringiensis effective against the Egyptian cotton leaf warm Spodoptera littoralis Boisd. Biotechnol. Lett., 2: 419-426.

Gómez I., Pardo-López L., Munoz-Garay C., Fernandez L.E., Pérez C., Sánchez J., Soberón M., and Bravo A. (2007) Role of receptor interaction in the mode of action of insecticidal Cry and Cyt toxins produced by Bacillus thuringiensis. Peptides 2 8: 1 6 9 -1 7 3.

Grochulski P., Masson L., Borisova S., Pusztai-Carey M., Schwartz J.-L., Brousseau R., and Cygler M. (1995) *Bacillus thuringiensis* CryIA(a) insecticidal toxin: crystal structure and channel formation. J. Mol. Biol., 254:447-464.

Guttmann D.M., and Ellar D.J. (2000) Phenotypic and genotypic comparisons of 23 strains from the Bacillus cereus complex for a selection of known and putative B. thuringiensis virulence factors. FEMS Microbiology Letters, 188: 7-13.

Haider M.Z., Knowles B.H., and Ellar D.J. (1986) Specificity of *Bacillus thuringiensis* var. *colmeri* insecticidal delta-endotoxin is determined by differential proteolytic processing of the protoxin by larval gut proteases. Eur. J. Biochem., 156: 531-540.

Hansen B.M., Damgaard P.H., Eilenberg J., and Pedersen J.C. (1998) Molecular and phenotypic characterization of *Bacillus thuringiensis* isolated from leaves and insects. J. Invertebr. Pathol., 71: 106-114.

Hansen B.M., and Hendriksen N.B. (2001) Detection of enterotoxic Bacillus cereus and Bacillus thuringiensis strains by PCR analysis. Applied and Environmental Microbiology.

Harms R.L., Martinez D.R. and Griego V.M. (1986) Isolation and Characterization of Coproporphyrin Produced by Four Subspecies of *Bacillus thuringiensis*. Appl Environ Microbiol., 51(3): 481-486.

Hofte H. and Whiteley H.R. (1989) Insecticidal Crystal Proteins of Bacilllus thuringiensis. Microbiological Reviews, 53(2): 242-255.

Ichimatsu T., Mizuki E., Nishimura K., Akao T., Saitoh H., Higuchi K., and Ohba M. (2000) Occurrence of *Bacillus thuringiensis* in fresh waters of Japan. *Curr. Microbiol.*, 40: 217-220.

Ito A., Sasaguri Y., Kitada S., Kusaka Y., Kuwano K., Masutomi K., Mizuki E., Akao T., and Ohba M. (2004) A Bacillus thuringiensis crystal protein with selective cytocidal action to human cells. J Biol. Chem., 279, 21282-21286.

Jin B., van Leeuwen H.J., Patel B., Doelle H.W., and Yu Q. (1999) Production of fungal protein and glucoamylase by *Rhizopus oligosporus* from starch processing wastewater. Process Biochemistry 34: 59–65.

Jin B., Yan X.Q., Yu Q., and van Leeuwen J.H. (2002) A comprehensive pilot plant system for fungal biomass protein production and wastewater reclamation. Advances in Environmental Research 6: 179-189.

Johnson D.E., and McGaughey W.H. (1996) Contribution of *Bacillus thuringiensis* spores to toxicity of purified Cry proteins towards Indianmeal moth larvae. *Curr Microbiol* 33, 54-59.

Johnson D.E., Oppert B., and McGaughey (1998) Spore Coat Protein Synergizes *Bacillus thuringiensis* Crystal Toxicity for the Indianmeal Moth (*Plodia interpunctella*). Current Microbiology 36: 278-282.

Joung K.-B., and Côté J.-C. (2001a) Phylogenetic analysis of Bacillus thuringiensis serovars based on 16S rRNA and gene restriction fragment length polymorphisms. Journal of Applied Microbiology, 90: 115-122.

Joung K.-B., and Côté J.-C. (2001b) A phylogenetic analysis of Bacillus thuringiensis serovars by RFLP-based ribotyping. Journal of Applied Microbiology, 91: 279-289.

Jung Y.-C, E. Mizuki E., Akao T., and Côté J.-C. (2007) Isolation and characterization of a novel Bacillus thuringiensis strain expressing a novel crystal protein with cytocidal activity against human cancer cells. J. Appl. Microbiol., 103: 65-79.

Jurat-Fuentes J.L., and Adang M.J. (2004) Characterization of a Cry1Ac-receptor alkaline phosphatase in susceptible and resistant Heliothis virescens larvae. Eur. J. Biochem., 271: 3127-3135.

Kang B.C., Lee S.Y., and Chang H.N. (1992) Enhanced spore production of *Bacillus thuringiensis* by fed batch culture. Biotechnology Letters, 14: 721-726.

Knight P., Crickmore N., and Ellar D.J. (1994) The receptor for Bacillus thuringiensis CryIA(c) delta-endotoxin in the brush border membrane of the lepidopteran Manduca sexta is aminopeptidase N. Mol. Microbiol., 11: 429-436.

Krieg A., Huger A.M., Langenbruch G.A., Schnetter W., 1983. *Bacillus thuringiensis* var. *tenebrionis*: a new pathotype effective against larvae of Coleoptera. Z. Angew. Entomol. 96, 500–508.

Kunitate, A., M. Okamoto, and I. Ohmori. 1989. Purification and characterization of a thermostable serine protease from *Bacillus thuringiensis*. Agric. Biol. Chem. 53:3251–3256.

Lachhab K., Tyagi R.D., and Valero J.R. (2001) Production of *Bacillus thuringiensis* biopesticides using wastewater sludge as a raw material: effect of inoculum and sludge solids concentration. Process Biochemistry, 37: 197-208.

LeBlanc M.E. (2004) Effets des différentes stratégies et pré-traitements des biosolides municipaux sur la croissance, la sporulation et l'entomotoxicité de *Bacillus thuringiensis var. kurstaki*. MSc thesis, INRS-ETE, Université du Quebec, Canada, 164 pp.

Lecadet M.M., Frachon E., Cosmao Dumanoir V., Ripouteau H., Hamson S., Laurent P., Thiéry I. (1999) Updating the H-antigen classification of *Bacillus thuringiensis*. J. Appl. Microbiol., 86: 660–672.

Lee D.H., Machii J., and Ohba M. (2002) High frequency of *Bacillus thuringiensis* in feces of herbivorous animals maintained in a zoological garden in Japan. Appl. Entomol. Zool., 37(4): 509-516.

Lee D.W., Akao T., Yamashita S., Katayama H., Maeda M., Saitoh H., Mizuki E., and Ohba M. (2000) Noninsecticidal parasporal proteins of a Bacillus thuringiensis serovar shandongiensis isolate exhibit a preferential cytotoxicity against human leukemic T cells. Biochem. Biophys. Res. Commun., 272: 218-223.

Lee, M. K., Walters, F. S., Hart, H., Palekar, N., and Chen, J. (2003) The mode of action of the *Bacillus thuringiensis* vegetative insecticidal protein Vip3A differs from that of Cry1Ab delta-endotoxin. Appl. Environ. Microbiol., 69, 4648–4657.

Lertcanawanichakul M. and Wiwat C. (2000) Improved shuttle vector for expression of chitinase gene in *Bacillus thuringiensis*. Letters in Applied Microbiology 31: 123-128.

Lertcanawanichakul M., Wiwat C., Bhumiratana A. and Dean D.H. (2004) Expression of chitinase-encoding genes in *Bacillus thuringiensis* and toxicity of engineered *B. thuringiensis* subsp. *aizawai* toward *Lymantria dispar* larvae. Current Microbiology 48: 175–181.

Levinson B.L., Kasyan K.J., Chiu S.S., Currier T.C. and Gonzalez, J.M.Jr. (1990) Identification of ,-Exotoxin Production, Plasmids Encoding β-Exotoxin, and a New Exotoxin in Bacillus thuringiensis by Using High-Performance Liquid Chromatography. Journal of Bacteriology, 172(6): 3172-3179.

Li E., and Yousten A.A. (1975) Metalloprotease from Bacillus thuringiensis. Applied Microbiology 30: 354-361.

Li J., Carroll J., and Ellar D.J. (1991) Crystal structure of insecticidal d-endotoxin from *Bacillus thuringiensis* at 2.5 Å resolution. Nature 353: 815-821.

Li R.S., Jarrett P. and Burges H.D. (1987) Importance of spores, crystals, and δ-endotoxins in the pathogenicity of different varieties of *Bacillus thuringiensis* in *Galleria mellonella* and *Pieris brassicae*. J Invertebr Pathol 50, 277-284.

Lin Y. and Xiong G. (2004) Molecular cloning and sequence analysis of the chitinase gene from *Bacillus thuringiensis* serovar *alesti*. Biotechnology Letters 26: 635–639.

Lisanky S.G., Quinlan R., and Tassoni G. (1993) The Bacillus thuringiensis production handbook. CPL Press, Newbury. UK.

Liu B.-L., Tzeng Y.M. (2000) Characterization Study of the Sporulation Kinetics of Bacillus thuringiensis. Biotechnol. Bioeng. 68: 11-17.

Liu M., Cai Q.X., Liu H.Z., Zhang B.H., Yan J.P. and Yuan Z.M. (2002) Chitinolytic activities in *Bacillus thuringiensis* and their synergistic effects on larvicidal activity. Journal of Applied Microbiology 93: 374–379.

Liu Y., Tabashnik B.E., Moar W.J., and Smith R.A. (1998) Synergism between *Bacillus thuringiensis* spores and toxins against resistant and susceptible diamond back moths (*Plutella xylostella*). Appl. Environ. Microbiol., 64:1385-1389.

Loguercio, L. L., Barreto, M. L., Rocha, T. L., Santos, C. G., Teixeira, F. F., and Paiva, E. (2002) Combined analysis of supernatant-based feeding bioassays and PCR as a firsttier screening strategy for Vip-derived activities in *Bacillus thuringiensis* strains effective against tropical fall armyworm. *J. Appl. Microbiol.*, **93**, 269–277.

Lövgren A., Zhang M., Engström A., Dalhammar G., and Landén R. (1990) Molecular characterization of immune inhibitor A, a secreted virulence protease from *Bacillus thuringiensis*. Mol. Microbiol., 4: 2137-2146.

MacIntosh, S. C., G. M. Kishore, F. J. Perlak, P. G. Marrone, T. B. Stone, S. R. Sims, and R. L. Fuchs. 1990. Potentiation of *Bacillus thuringiensis* insecticidal activity by serine protease inhibitors. J. Agric. Food Chem. 38:1145–1152.

Malladi B., and Ingham S.C. (1993) Thermophilic aerobic treatment of potato-processing wastewater. World Journal of Microbiology and Biotechnology 9: 45-49.

Manker D.C., Lidster W.D., Starnes R.L. and MacIntosh S.C. (1994) Potentiator of *Bacillus* pesticidal activity. Patent Coop. Treaty, WO94/09630.

Manker, D. C., Lidster, W. D., MacIntosh, S. C., and Starnes, R. L. (2002) Potentiator of Bacillus pesticidal activity. United States Patent 6,406,691.

Meadows M.P., Ellis D.J., Butt J., Jarret P., and Burges H.D. (1992) Distribution, frequency, and diversity of *Bacillus thuringiensis* in animal feed mill. Appl. Environ. Microbiol., 58:1344-1350.

Mizuki E., Ohba M., Akao T., Yamashita S., Saitoh H. and Park Y.S. (1999) Unique activity associated with noninsecticidal Bacillus thuringiensis parasporal inclusions: in vitro cell-killing action on human cancer cells. J. Appl. Microbiol., 86: 477-486.

Mizuki E., Park Y.S., Saitoh H., Yamashita S., Akao T., Higuchi K., and Ohba M. (2000) Parasporin, a human leukemic cell-recognizing parasporal protein of *Bacillus thuringiensis*. *Clin. Diagn. Lab. Immunol.* **7**, 625–634.

Mizuki E., Maeda M., Tanaka R., Lee D.W., Hara M., Akao T., Yamashita S., Kim H.S., Ichimatsu T., and Ohba M. (2001) *Bacillus thuringiensis*: a common member of microflora in activated sludges of a sewage treatment plant. *Curr. Microbiol.* 42: 422-425.

Mohammedi S., Bala Subramanian S., Yan S., Tyagi R.D., and Valéro J.R. (2006) Molecular screening of Bacillus thuringiensis strains from wastewater sludge for biopesticide production. Process Biochemistry, 41: 829-835.

Morris O.N., Converse V., and Kanagaratnam P., and Davies J.S. (1996) Effect of Cultural Conditions on Spore–Crystal Yield and Toxicity of *Bacillus thuringiensis* subsp. *aizawai* (HD133). Journal of Invertebrate Pathology, 67: 129–136

Nair, J. R., Narasimman, G., and Sekar, V. (2004) Cloning and partial characterization of zwittermicin A resistance gene cluster from *Bacillus thuringiensis* subsp. *kurstaki* strain HD1. *J. Appl. Microbiol.*, **97**, 495–503.

Nguyen Cong Hao, Nguyen Cuu Thi Huong Giang, Nguyen Cuu Khoa, Nguyen Thanh Son (1996) Synthesis and application of insect attractants in Vietnam. Resources, Conservation and Recycling 18: 59-68.

Nickerson K.W., and Bulla L.A.Jr. (1974) Physiology of Sporeforming Bacteria Associated with Insects: Minimal Nutritional Requirements for Growth, Sporulation, and Parasporal Crystal Formation of Bacillus thuringiensis.Applied Microbiology, 28(1): 124-128.

Nickerson, K. W., and J. D. Swanson (1981) Removal of contaminating proteases from *Bacillus thuringiensis* parasporal crystals by density gradient centrifugation in NaBr. Eur. J. Appl. Microbiol. Biotechnol. 13:213–215.

Ogiwara K., Indrasith L.S., Asano S., and Hori H. (1992) Processing of delta-endotoxin from *Bacillus thuringiensis* subsp. *kurstaki* HD-1 and HD-73 by gut juices of various insect larvae. J. Invertebr. Pathol., 60: 121-126.

Porcar M., and Juáréz-Pérez V. (2003) PCR-based identification of Bacillus thuringiensis pesticidal crystal genes. FEMS Microbiology Reviews 26(2003) 419-432.

Rajasimman M., and Karthikeyan C. (2007) Aerobic digestion of starch wastewater in a fluidized bed bioreactor with low density biomass support. Journal of Hazardous Materials 143: 82-86.

Rajbhandari B.K., and Annachhatre A.P. (2004) Anaerobic ponds treatment of starch wastewater: case study in Thailand. Bioresource Technology 95: 135–143.

Regev A., Keller M., Strizhov N., Sneh B., Prudovsky E., Chet I., Ginzberg I., Koncz-Kalman Z., Koncz C., Schell J. and Zilberstein A. (1996) Synergistic activity of a *Bacillus thuringiensis* δ-endotoxin and a bacterial endochitinase against *Spodoptera littoralis*. Applied and Environmental Microbiology 62: 3581–3586.

Rojas-Avelizapa L. I., Cruz-Camarillo R., Guerrero M. I., Rodríguez-Vázquez R. and Ibarra J. E. (1999) Selection and characterization of a proteochitinolytic strain of *Bacillus thuringiensis*, able to grow in shrimp waste media. World Journal of Microbiology and Biotechnology 15: 299–308.

Salamitou S., Ramisse F., Brehelin M., Bourget D., Gilois N., Gominet N. et al. (2000) The *plcR* regulon is involved in the opportunistic properties of *Bacillus thuringiensis* and *Bacillus*
cereus in mice and insects. Microbiol., 146: 2825-2832.

Sampson M.N. and Gooday G.W. (1998) Involvement of chitinases of *Bacillus thuringiensis* during pathogenesis in insects. Microbiology 144: 2189–2194.

Scherrer P., Luthy P., and Trumpi P. (1973) Production of δ-Endotoxin by Bacillus thuringiensis as a Function of Glucose Concentrations. Applied Microbiology, 25: 644-646.

Schesser J.H., and Bulla L.A.Jr. (1978) Toxicity of Bacillus thuringiensis Spores to the Tobacco Hornworm, Manduca sexta. Applied and Environmental Mocrobiology, 35(1): 121-123.

Schnepf H.E., Whiteley H.R., 1981. Cloning and expression of the *Bacillus thuringiensis* crystal protein gene in *Escherichia coli*. Proc. Natl. Acad. Sci. USA., 78: 2893-2897.

Schnepf E., Crickmore N., van Rie J., Lereclus D., Baum J., Feitelson J., Zeigler D.R., and Dean D.H. (1998) *Bacillus thuringiensis* and its pesticidal crystal proteins. Microbiology and Molecular Biology Reviews, 62(3): 775-806.

Selvapandiyan, A., Arora, N., Rajagopal, R., Jalali, S. K., Venkatesan, T., Singh, S. P., and Bhatnagar, R. K. (2001) Toxicity analysis of N- and C-terminus-deleted vegetative insecticidal protein from *Bacillus thuringiensis*. *Appl. Environ. Microbiol.*, 67, 5855–5858.

Siden I., Dalhammar G., Telander B., Boman H.G., and Somerville H. (1979) Virulence factors in *Bacillus thuringiensis*: purification and properties of a protein inhibitor of immunity in insects. J. Gen. Microbiol., 114: 45-52.

Siegel J.P. (2001) The Mammalian Safety of *Bacillus thuringiensis*-Based Insecticides. Journal of Invertebrate Pathology, **77** : 13-21.

Silo-Suh, L. A., Stabb, E. V., Raffel, S., and Handelsman, J. (1998) Target range of zwittermicin A an aminopolyol antibiotic from *Bacillus cereus*. *Curr. Microbiol.*, **37**, 6–11.

Sirichotpakorn N., Rongnoparut P., Choosang K., and Panbangred W. (2001) Coexpression of chitinase and the *cry11Aa1* toxin genes in *Bacillus thuringiensis* serovar *israelensis*. Journal of Invertebrate Pathology 78: 160–169.

Smirnoff W.A. et Valéro J.R. (1972) Perturbations métaboliques chez *Choristoneura fumiferana* au cours de l'infection par *Bacillus thuringiensis* seul ou en présence de chitinase. Rev. Can. de Biol. 31 : 163-169.

Smirnoff W.A. (1973) Results of tests with *Bacillus thuringiensis* and chitinase on larvae of the spruce budworm. Journal of Invertebrate Pathology 21:116–118.

Smirnoff W.A. (1974) Three years of aerial field experiments with *Bacillus thuringiensis* plus chitinase formulation against the spruce bud worm. Journal of Invertebrate Pathology 24: 344–348.

Smith R.A., and Couche G.A. (1991) The philloplane as a source of *Bacillus thuringiensis* variants. Appl. Environ. Microbiol. 57:311–331.

Stabb E.V., Jacobsen L.M. and Handelsman J. (1994) Zwittermicin A-producing strains of *Bacillus cereus* from diverse soils. Appl. Environ. Microbiol., 60: 4404-4412.

Stohl, E. A., Milner, J. L., and Handelsman, J. (1999) Zwittermicin A biosynthetic cluster. *Gene*, **237**, 403–411.

Taguchi R., Asahi Y., and Ikezawa H. (1980) Purification and properties of phosphatidylinositol-specific phospholipase C of Bacillus thuringiensis. Biochim Biophys Acta, 619(1): 48-57.

Tan Y., and Donovan W.P. (2000) Deletion of *aprA* and *nprA* genes for alkaline protease A and neutral protease A from *Bacillus thuringiensis*: effect on insecticidal crystal proteins. Journal of Biotechnology, 84: 67-72.

Tantimavanich S., Pantowatana S., Bhumiratana A. and Panbangred W. (1997) Cloning of a chitinase gene into *Bacillus thuringiensis* subsp. *aizawai* for enhanced insecticidal activity. Journal of General Applied Microbiology 43:31–37.

Thamthiankul S., Suan-Ngay S., Tantimavanich S. and Panbangred W. (2001) Chitinase from *Bacillus thuringiensis* subsp. *pakistani*. Applied Microbiology and Biotechnology 56:395–401.

Thamthiankul S., Moar W.J., Miller M.E. and Panbangred W. (2004) Improving the insecticidal activity of *Bacillus thuringiensis* subsp.

aizawai against Spodoptera exigua by chromosomal expression of a chitinase gene. Applied Microbiology and Biotechnology 65: 183–192.

Tojo A., Samasanti W., Yoshida N., and Aizawa K. (1986) Effects of the three proteases from gut juice of the silkworm, *Bombyx mori*, on the two morphologically different inclusions of delta-endotoxin produced by *Bacillus thuringiensis* kurstaki HD-1 strain. Agric. Biol. Chem., 50: 575-580.

Tyagi R.D., Sikati Foko V., Barnabe S., Vidyarthi A.S., Valéro J.R., and Surampalli R.Y. (2002) Simultaneous production of biopesticide and alkaline proteases by *Bacillus thuringiensis* using sewage sludge as a raw material. Water Science and Technology 46: 247-254.

Vadlamudi R.K., Weber E., Ji I., Ji T.H., and Bulla L.A. Jr. (1995) Cloning and expression of a receptor for an insecticidal toxin of Bacillus thuringiensis. J. Biol. Chem., 270: 5490-5494.

Valaitis A.P., Jenkins J.L., Lee M.K., Dean D.H., and Garner K.J. (2001) Isolation and partial characterization of Gypsy moth BTR-270 an anionic brush border membrane glycoconjugate that binds Bacillus thuringiensis Cry1A toxins with high affinity. Arch. Ins. Biochem. Physiol. 46, 186–200.

Verma M., Brar S.K., Tyagi R.Y., Surampalli R.Y. and Valéro, J.R. (2007) Starch industry wastewater as a substrate for antagonist, *Trichoderma viride* production. Bioresource Technology Bioresource Technology 98: 2154-2162.

Vidyarthi A.S., Tyagi R.D., Valéro J.R., and Surampalli R.Y. 2002 Studies on the production of Bacillus thuringiensis based biopesticides using wastewater sludge as a raw material. Water Research, 36: 4850-4860.

Vilas-Bôas G.T. and Franco Lemos M.V. (2004) Diversity of *cry* genes and genetic characterization of *Bacillus thuringiensis* isolated from Brazil. Can. J. Microbiol., 50: 605-613.

Vu K.D., Tyagi R.D., Valéro J.R., Surampalli R.Y. (2008) Impact of different pH control agents on biopesticidal activity of *Bacillus thuringiensis* during the fermentation of starch industry wastewater. *Bioprocess Biosyst. Eng., DOI 10.1007/s00449-008-0271-z.* (online)

Vu K.D., Tyagi R.D., Brar S.K., Valéro J.R., Surampalli R.Y (2009a) Starch industry wastewater for production of biopesticides – Ramifications of solids concentrations. *Environmental Technology* 30(4): 393-405.

Vu K.D., Adjallé K.D., Tyagi R.D., Valéro J.R., Surampalli R.Y. (2009b) Bacillus thuringiensis based-biopesticides production using starch industry wastewater fortified with different carbon/nitrogen sources as fermentation media. *Manuscript to be submitted.*

Vu K.D., Tyagi R.D., Valéro J.R., Surampalli R.Y. (2009c). Fed-batch fermentation of *Bacillus thuringiensis* using starch industry wastewater as fermentation and fed substrate. Bioprocess Biosyst. Eng. 33: 691-700.

Vu K.D., Yan S., Tyagi R.D., Valéro J.R., Surampalli R.Y. (2009d) Induced production of chitinase to enhance entomotoxicity of *Bacillus thuringiensis* employing starch industry wastewater as a substrate. Bioresource Technology, 100: 5260-5269.

Vu K.D., Adjallé K.D., Tyagi R.D., Valéro J.R., Surampalli R.Y. (2009e) Recovery of chitinases and Zwittermicin A from *Bacillus thuringiensis* fermented broth to produce biopesticide with high biopesticidal activity. *Manuscript to be submitted.*

Vu K.D., Tyagi R.D., Valéro J.R., Surampalli R.Y. (2009f) Mathematical relationships between spore concentrations, delta-endotoxin levels, and entomotoxicity of Bacillus thuringiensis preparations produced in different fermentation media. Bioresource Technology 123: 303-311.

Warren G.W., Koziel M.G., Mullins M.A. (1998) Auxiliary proteins for enhancing the insecticidal activity of pesticidal proteins. Novartis, US patent 5770696

Whalon M.E., and Wingerd B.A. (2003) Bt: Mode of action and use. Archives of Insect Biochemistry and Physiology 54:200-211.

Whiteley H.R., and Schnepf E. (1986) The molecular biology of parasporal crystal body formation in Bacillus thuringiensis. Ann. Rev. Microbiol., 40: 549-576.

WHO (1999) Guidelines specification for bacterial larvicides for public health use. In: WHO/CDS/CPC/WHOPES/99.2 (ed) Report of the WHO informal consultation, 28–30 April 1999. Geneva: World Health Organization.

Wiwat C., Lertcanawanichakul M., Siwayapram P., Pantuwatana S. and Bhumiratana A. (1996) Expression of chitinase encoding genes from *Aeromonas hydrophila* and *Pseudomonas hydrophila* in *Bacillus thuringiensis* subsp. *israelensis*. Gene 179: 119 –126.

Wiwat C., Thaithanum S., Pantuwatana S. and Bhumiratana A. (2000) Toxicity of chitinase-producing *Bacillus thuringiensis* ssp. *kurstaki* HD-1 (G) toward *Plutella xylostella*. Journal of Invertebrate Pathology 76: 270–277.

Yasutake K., Ngo D.B., Kagoshima K., Uemori A., Ohgushi A., Maeda M., Mizuki E., Yu Y.M., and Ohba M. (2005) Occurrence of parasporin-producing *Bacillus thuringiensis* in Vietnam. Can. J. Microbiol. 52: 365-372.

Yezza A., Tyagi R.D., Valéro J.R., and Surampalli R.Y. (2005a) Production of *Bacillus thuringiensis* based biopesticides by batch and fed-batch culture using wastewater sludge as a raw material. Journal of Chemical Technology and Biotechnology 80: 502-510.

Yezza A., Tyagi R.D., Valéro J.R., and Surampalli R.Y. (2005b) Wastewater sludge pre-treatment for enhancing entomotoxicity produced by Bacillus thuringiensis var. kurstaki. World Journal of Microbiology & Biotechnology, 21: 1165-1174.

Yezza A., Tyagi R.D., Valéro J.R., and Surampalli R.Y. (2005c) Influence of pH control agents on entomotoxicity potency of Bacillus thuringiensis using different raw materials. World Journal of Microbiology & Biotechnology, 21: 1549–1558.

Yezza A., Tyagi R.D., Valéro J.R., and Surampalli R.Y. (2006) Bioconversion of industrial wastewater and wastewater sludge into *Bacillus thuringiensis* based biopesticides in pilot fermentor. Bioresource Technology, 97: 1850-1857.

Yu, C. G., Mullis, M. A., Warren, G. W., Koziel, M. G., and Estruch, J. J. (1997) The *Bacillus thuringiensis* vegetative insecticidal protein Vip3A lyses midgut epithelium cells of susceptible insects. *Appl. Environ. Microbiol.*, **63**, 532–536.

Zhao C, Luo Y., Song C., Liu Z., Chen S., Yu Z., and Sun M. (2007) Identification of three Zwittermicin A biosynthesis-related genes from *Bacillus thuringiensis* subsp. *kurstaki* strain YBT-1520. Arch Microbiol (2007) 187:313–319.

Zhong W.F., Jiang L.H., Yan W.Z., Cai P.Z., Zhang Z.X. and Pei Y. (2003) Cloning and sequencing of chitinase gene from *Bacillus thuringiensis* serovar *israelensis*. Acta Genetica Sinica 30: 364-369.

Zhong W.F., Fang J.C., Cai P.Z., Yan W.Z., Wu J. and Guo H.F. (2005) Cloning of the *Bacillus thuringiensis* serovar *sotto* chitinase (*Schi*) gene and characterization of its protein. Genetics and Molecular Biology 28 (4): 821-826.

Zouari N., and Jaoua S. (1999a) Production and characterization of metalloproteases synthesized concomitantly with δ-endotoxin by *Bacillus thuringiensis* subsp. *kurstaki* strain grown on gruel-based media. Enzyme and Microbial Technology, 25 364-371.

Zouari N., and Jaoua S. (1999b) The effect of complex carbon and nitrogen, salt, Tween-80 and acetate on delta-endotoxin production by a Bacillus thuringiensis subsp kurstaki. Journal of Industrial Microbiology & Biotechnology, 23: 497-502.

CHAPITRE 2. STRATÉGIES POUR L'AMÉLIORATION D'ENTOMOTOXICITÉ (TX) DE BIOPESTICIDE À BASE DE B. THURINGIENSIS EN UTILISANT LES EAUX USÉES D'INDUSTRIE D'AMIDON (SIW) COMME MATIÈRES PREMIÈRES

PARTIE 2.1. Impact de différents agents de contrôle du pH sur l'activité biopesticide de *Bacillus thuringiensis* (Btk HD-1) produit par bioréaction en utilisant les eaux usées d'industrie d'amidon comme substrats de fermentation

Résumé

Différents agents de contrôle du pH ($NaOH/H_2SO_4$—SodSulp, $NaOH/CH_3COOH$—SodAcet, NH_4OH/CH_3COOH—AmmoAcet et NH_4OH/H_2SO_4—AmmoSulp) ont été utilisés pour maintenir le pH à 7 au cours de la production de Btk HD-1 par fermentation en utilisant les eaux usées d'industrie d'amidon comme matières premières. Il a été constaté que AmmoSulp et SodSulp sont respectivement les meilleurs agents pour contrôler le pH lors de la production de protéases alcalines et d'amylases, alors que le bouillon fermenté obtenu en utilisant SodAcet comme agents tampons contenait la plus forte concentration de delta-endotoxines (1043,0 mg/L) et la plus haute valeur entomotoxique (Tx-18,4 x 10^9 SBU/l). La Tx de la suspension du culot (à un dixième du volume initial) avec le SodAcet comme agents de contrôle du pH atteignait 26,7 x 10^9 SBU/l soit la valeur de Tx la plus élevée par rapport à trois autres agents de contrôle du pH.

Impact of different pH control agents on biopesticidal activity of *Bacillus thuringiensis* during the fermentation of starch industry wastewater

Abstract

Different pH control agents ($NaOH/H_2SO_4$ - SodSulp, $NaOH/CH_3COOH$ - SodAcet, NH_4OH/CH_3COOH – AmmoAcet and NH_4OH/H_2SO_4 - AmmoSulp) were used to investigate their effects on growth, enzymes production (alkaline protease and amylase), and entomotoxicity of *Bacillus thuringiensis* var. *kurstaki* HD-1 (Btk) against eastern spruce budworm larvae (*Choristoneura fumiferana*) using starch industry wastewater (SIW) as a raw material in a 15 L fermentor. AmmoSulp and SodSulp were found to be the best pH control agents for alkaline protease and amylase production, respectively; whereas, the fermented broth obtained by using SodAcet as pH control agents recorded the highest delta-endotoxin production of 1043.0 mg/L and entomotoxicity value 18.4 x 10^9 SBU/L. Entomotoxicity of re-suspended centrifuged pellet in one tenth of original volume in case of SodAcet as pH control agents was 26.7 x 10^9 SBU/L and was the highest value compared to three other pH control agents.

Keywords: Starch industry wastewater; *Bacillus thuringiensis* var. *kurstaki*; Entomotoxicity; Biopesticides; Bioconversion

2.1.1. Introduction

Microorganisms require nutrients and minerals for their growth and metabolic product formation and the requirements of nutrient and minerals vary with the type of microorganism as well as with the nature of nutrient medium under investigation [1, 2]. Starch industry wastewater (SIW) contains most of the important ingredients (such as residual starch, reducing sugars, soluble proteins, minerals) required for growth of microorganisms [3] and is also a suitable medium for production of many useful products such as microbial biomass protein, fungal α-amylase, lactic acid and biopesticides [3-6]. The bioconversion of SIW into useful products has many important considerations: (1) low production cost of value added products; (2) environment friendly product; 3) minimization of wastewater treatment cost; 4) sequestration of carbon from wastewater and hence reduction of greenhouse gas generation; 5) reduction of adverse environmental impact. Bioconversion of SIW into *Bacillus thuringiensis* (Bt) based biopesticides has been studied in our lab and the results obtained did establish that the biopesticidal activity or entomotoxicity (Tx) of Bt fermented SIW was higher than that observed in synthetic medium, slaughter house wastewater and different wastewater sludges [6, 7].

During the production of Bt based biopesticides, delta-endotoxin synthesis and entomotoxic potential may be affected by the nitrogen source available in the medium. In this regard, Avignone Rossa et al. [8] investigated the influence of different combinations of organic and inorganic nitrogen sources and C: N ratios on growth and delta-endotoxin production by Bt var. *israelensis*. These authors found that delta-endotoxin concentration increased when the medium was supplemented with $(NH_4)_2SO_4$. However, the delta-endotoxin: biomass dry weight ratio was unaffected by different C:N ratios. They also suggested that yeast extract as organic nitrogen source could be partially replaced by $(NH_4)_2SO_4$ with a significant increase in delta-endotoxin production. Kraemer-Schafhalter and Moser [9] concluded that NH_4^+ ion used as an inorganic nitrogen source was necessary for synthesis of intracellular protein, cell wall components and delta-endotoxin.

It has been reported that vegetative growth of *Bacillus* sp. causes a build up of acetic and pyruvic acids (intermediate products of glucose metabolism) which are later used during the onset of sporulation [9-11]. In glucose rich medium as well as in a medium containing complex source of carbohydrates (wastewaters, wastewater sludge), Bt utilizes glucose and/or easily biodegradable organic mattter as primary carbon source producing organic acids (lactate, pyruvate, and acetate). These products cause the medium pH to decrease and become acidic. When glucose or easily biodegradable carbon source is depleted, the produced acids or any other complex protein source (complex medium like waste materials) in the medium are then consumed as secondary carbon sources. Hence, pH value of the broth increases [12, 13]. Therefore, the control of pH during Bt fermentation is essential.

On the other hand, the pH control agents can serve as a useful tool to provide extra nutrients (nitrogen and carbon sources), that may not be present in sufficient quantity in the medium, required for Bt growth and toxin synthesis, especially, when waste materials are used as source of nutrients. For example, Yezza et al. [13] found that replacing conventional pH control agents ($NaOH/H_2SO_4$) with NH_4OH/CH_3COOH as pH control agents during Bt fermentation employing wastewater sludge as a raw material, the Tx value and protease activity (PA) were enhanced by 22% and 14%, respectively whereas in the case of semi-synthetic soy medium Tx and PA were increased by 12% and 53%, respectively. However, the research was limited to replace $NaOH/H_2SO_4$ with CH_3COOH/NH_4OH as pH control agents during Btk fermentation of the wastewater sludge and the synthetic medium.

Thus, when SIW is employed as a raw material for Bt growth, it is not clear: 1) if replacement of conventional pH control agents will augment the entomotoxic potential and; 2) if both the conventional pH control agents or one of the agent should be replaced with CH_3COOH (carbon source) and NH_4OH (nitrogen source). Or, in other words, if one of the nutrients (C and N) is deficient or both of them are in short supply in SIW. Moreover, compositions of wastewater sludge (used in earlier studies) and SIW are entirely different not only in terms of concentration of different components but also in terms of their availability for growth, spores production and insecticidal proteins synthesis by Bt.

Therefore, the objective of this research was to investigate the impact of replacing conventional pH control agents ($NaOH/H_2SO_4$) with C and N furnishing pH control agents on growth, delta-endotoxin synthesis, enzyme production (alkaline protease, amylase) and entomotoxic potential of Bt while using SIW as raw material. The conventional pH control agents ($NaOH/H_2SO_4$ - SodSulp) were replaced with three different combinations of pH control agents: $NaOH/CH_3COOH$ (SodAcet), NH_4OH/H_2SO_4 (AmmoSulp) and NH_4OH/CH_3COOH (AmmoAcet) to fortify extra source of carbon and nitrogen.

2.1.2. Materials and Methods

2.1.2.1. Bacterial strain and inoculum preparation

Bacillus thuringiensis var. *kurstaki* HD-1 (ATCC 33679) (Btk) was used in this study. The Btk was subcultured and streaked on tryptic soya agar (TSA) plates, incubated for 48 h at 30 ± 0.1 °C and then preserved at 4 ± 0.1 °C for future use.

All media used for inoculum preparation were adjusted to pH 7.0 ± 0.1 before autoclaving. A loopful of Btk grown on TSA plate was used to inoculate a 500-ml Erlenmeyer flask containing 100 ml of sterilised tryptic soya broth (TSB) medium. The flask was incubated in a rotary shaker at 300 revolutions per min (rpm) and at 30 ± 0.1 °C for 8–12 h. A 2% (v/v) inoculum from this flask was then used to inoculate 500-ml Erlenmeyer flasks containing 100 ml of sterilised SIW. These flasks were incubated for 8-12 h. The actively growing cells from these flasks were used as a pre-culture for the production of Btk based biopesticides in fermentor [13].

2.1.2.2. Btk production medium

During the course of this study, SIW from ADM-Ogilvie (Candiac, Québec, Canada) was used as a raw material for Btk growth. SIW was sampled and stored at 4 ± 0.1 °C to avoid its deterioration due to slow growth of microorganisms. Total solids (TS), volatile solids (VS), suspended solids (SS) and volatile suspended solids (VSS), pH, total carbon (C_t), total nitrogen (N_t), ammonia nitrogen ($N–NH_3$), and total phosphorus (P_t) were determined according to the Standard Methods [14]. Different physico-chemical characteristics of SIW are presented in Table 6.

2.1.2.3. Fermentation procedure with different pH control agents

Fermentation was carried out in a stirred tank 15 L bioreactor (working volume: 10 L, Biogenie, Que., Canada) equipped with accessories and programmable logic control (PLC) system for dissolved oxygen (DO), pH, anti-foam, impeller speed, aeration rate and temperature. The software (iFix 3.5, Intellution, USA) allowed automatic set-point control and integration of all parameters via PLC.

Tableau 6. Characteristics of starch industry wastewater (SIW)

Parameter	Concentration (g/L)*
Total solids [TS]	13.5 ± 0.6 (g/L)
Volatile solids [VS]	11.2 ± 0.4 (g/L)
Suspended solids [SS]	4.4 ± 0.2 (g/L)
Volatile suspended solids [VSS]	4.2 ± 0.2 (g/L)
pH	3.4 ± 0.2
	Concentration (g/kg TS)*
Ct	502 ± 8
N_t	39.0 ± 1.0
$N\text{-}NH_3$	1.98 ± 0.01
P_t	10.7 ± 0.2

* The presented values are the mean ± standard deviation (SD).

Before each sterilisation cycle, polarographic pH-electrode (Mettler Toledo, USA) was calibrated using buffers of pH 4 and 7 (VWRCanada). The oxygen probe (Mettler Toledo, USA) was calibrated to zero (using N_2 degassed water) and 100% (air saturated water). Subsequently, fermenter was charged with starch industry wastewater (10 L) and polypropylene glycol (PPG, Sigma-Canada) (10 mL, 0.1%, v/v) solution as an anti-foam agent. The fermenter with medium was sterilised in situ at 121 °C for 30 min. When the fermenter cooled down to 30 ± 0.1 °C, DO probe was recalibrated to zero by sparging N_2 gas and 100% saturation by sparging air at agitation rate of 500 rpm. The fermenter was then aseptically inoculated (2% v/v inoculum) with preculture of Btk in exponential phase (8-12 h old). In order to keep the DO above 25% saturation, air flow rate and agitation rate were varied between

2-2.5 Litre of air per minute (LPM) and 300–350 revolution per minute (rpm), respectively. This ensured the critical DO level for Bt above 25% [15]. The temperature was maintained at 30 ± 0.1 °C by circulating water through the fermenter jacket. Fermentation pH was controlled automatically at 7.0 ± 0.1 through computer-controlled peristaltic pumps by addition of four different combination of pH control agents: SodSulp; AmmoAcet; AmmoSulp and SodAcet. The strength of NaOH, H_2SO_4, NH_4OH and CH_3COOH solutions used for pH control were 4M, 3M, 30% (v/v) and 10M, respectively. Dissolved oxygen and pH were continuously monitored.

Samples were collected at regular intervals to monitor the changes in total cell count, spore count and Tx. A part of all collected samples during fermentation was centrifuged at 9000 *g* for 20 min at 4 °C. The supernatant thus obtained was used to determine different enzyme activities (alkaline protease, amylase). At the end of fermentation, fermented broth obtained from different experiments conducted using different pH control agents were adjusted to pH 4.5 and centrifuged at 9000 g for 30 min [6]. The obtained pellets (spore and delta-endotoxin complex) were re-suspended with supernatant to obtain one-tenth volume of original fermented broth. The suspended pellets were subjected to bioassay.

2.1.2.4. Protease and amylase activity assay

Protease activity (PA) was determined according to modified Kunitz (1947) method [17]. A part of the supernatant (in triplicate) was appropriately diluted with borate buffer pH 8.2 ± 0.1. PA was assayed by incubating 1 ml enzyme solution with 5 ml casein for 10 min at 37 °C in a constant-temperature water-bath. The reaction was terminated by adding 5 ml 10% trichloroacetic acid (TCA). This mixture was incubated for 30 min to precipitate the total non-hydrolyzed casein. Parallel blanks were prepared with inactivated casein. At the end of the incubation period, samples and blanks were filtered through Whatman 934-AH paper. The absorbance of the filtrate was measured at 275 nm. One PA unit was defined as the amount of enzyme preparation required to liberate 1 μmol (181 μg) tyrosine from casein per minute in pH 8.2 buffer at 37°C. The presented values are the mean of three determination of two separate experiments ± Standard Deviation (SD).

Amylolytic activity (AA) was measured according to the method of Miller [18, 19]. One percent starch (0.5 ml) in Na-acetate 50 mM buffer at pH 6 was mixed with 0.5 ml of appropriately diluted enzyme sample and incubated for 30 min at 40 ± 0.1 °C. The reaction was stopped with 3 ml DNS (3,5-dinitrosalicylic acid) reagent and boiled for 5 min. Twenty mL of distilled water was added to the reaction medium, and the absorbance was determined at 545 nm. Parallel blanks were run with inactivated enzyme samples treated for 5 min in boiling water. One AA unit was defined as the amount of enzyme required to liberate 1 mg of glucose/min under the experimental conditions. The presented values are the mean of three repeated samples analysis of two separate experiments ± SD.

2.1.2.5. Estimation of total cell count (TC) and spore count (SC)

The TC and SC were performed by counting colonies grown on TSA medium [13]. For all counts, the average of at least triplicate plates was used for each tested dilution. For enumeration, 30–300 colonies were enumerated per plate. The results were expressed as colony forming units per mL (CFU/ml).

2.1.2.6. Estimation of delta-endotoxin production

Delta-endotoxin concentration was determined based on the solubilisation of insecticidal crystal proteins in alkaline condition [19-21]: 1 ml of each samples collected at the end of fermentation (48h) were centrifuged at 10000 g for 10 min at 4°C. The pellet containing a mixture of spores, insecticidal crystal proteins, cell debris and residual suspended solids was used to estimate the concentration of alkali soluble insecticidal crystal proteins (delta-endotoxin). These pellets were washed three times with 1 ml of 0.14 M NaCI - 0.0 1 % Triton X- 100 solutions. The washing helped in eliminating the soluble proteins and proteases that might be adhered on to the pellet and could affect the integrity of the crystal protein. The insecticidal crystal proteins in the pellet were dissolved with 0.05N NaOH (pH 12.5) for three hours at 30°C with stirring. The suspension was centrifuged at 10000 g for 10 min. at 4°C and the pellet, containing spores, cell debris and residual suspended solids was discarded. The supernatant, containing the alkali soluble insecticidal crystal proteins was used to determine the delta-endotoxin concentration by Bradford method using bovine serum albumin as standard protein [22]. The presented values were the mean of three determination of two separate experiments ± SD.

2.1.2.7. Bioassay technique:

Entomotoxicity (Tx) of Btk was estimated by bioassay against third instar larvae of spruce budworm (Lepidoptera: *Choristoreuna fumiferena*) following the diet incorporation method of Dulmage et al. and Beegle [23, 24]. The commercial preparation 76B FORAY (Valent Biosciences Corporation Libertyville, IL) was used as a reference standard to analyse the Tx. The detail procedure of bioassay technique was presented in Yezza et al. [13]. The Tx was evaluated by comparing the mortality percentages of dilutions of the samples (suspended pellet as well as whole fermented broth) with same dilutions of the standard. In our research, Tx was expressed as relative spruce budworm units (SBU/L). On comparison of Tx of Bt fermented samples, it was found that SBU reported in our study was 20–25% higher than IU [13]. The presented values are the mean of three determination of two independent experiments ± SD.

2.1.3. Results and discussion

2.1.3.1. Effect of different pH control agents on the growth and sporulation of Btk

2.1.3.1.1. Profiles of total cell count and spore count

Temporal profiles of total cell count (TC) and spore count (SC) during Btk fermentation of SIW with different pH control agents are presented in Fig. 1a and the maximum values of maximum specific growth rate (μ_{max}), TC, SC and % sporulation (spore count/total cell count) are presented in Table 7. In general, use of NH_4OH in combination with either H_2SO_4 or CH_3COOH tend to decrease the maximum values of μ_{max}, TC, VS and maximum sporulation at the end of fermentation (Table 7). This suggested inhibitory effects of NH_4OH on the overall performance of the process. The maximum total cell count of Btk occurred during 9-12h of fermentation process (Table 7). Irrespective of different pH control agents, the maximum spore count occurred at 24 h of fermentation followed by a slight decrease (Table 7). A decrease in spore count between 24-36 h (Fig. 8a) could be due to the fact that some of the completely sporulated cells had been formed during the early stage of fermentation process. These cells lysed and released spores, parasporal crystals and other intracellular material into the medium [25]; if appropriately stimulated, the released spores could initiate germination leading to vegetative growth [26]. Even in case where the required nutrients in the medium were exhausted, the spores could germinate due to the intracellular material (released in the medium during spore release) which served as a source of nutrients. The vegetative cells produced as a results of spore germination would have entered into the sporulation process again after depletion of released intracellular nutrients.

2.1.3.1.2 Quantity of pH control agents used during fermentation processes

In general, during the fermentation a certain amount of minerals/nutrients was provided in the form of Na^+, NH_4^+, SO_4^{2-} and CH_3COO^- through the use of different pH control agents. These additional minerals/nutrients also influenced growth, enzyme production and entomotoxicity of Btk depending on the available nutrients in SIW. As stated above, a decrease in maximum specific growth rate (μ_{max}) and the value of other parameters associated with growth of Btk (Table 7) could be due to the addition of high concentration NH_4^+ ions.
When SodSulp and SodAcet were used as pH control agents, the amount of Na^+ used was 2.30 and 3.22 g in 10 L working volume of bioreactor, respectively, whereas the amount of NH_4^+ used in case of AmmoSulp and AmmoAcet as pH control agents was 5.40 and 4.50 g in 10 L working volume of bioreactor, respectively (Table 8). Thus, the amount of NH_4^+ added to control pH was higher as compared to the amount of Na^+. In fact, the results in Table 7 suggested that spore concentration and sporulation at the end of fermentation using AmmoSulp and AmmoAcet as pH control agents was significantly lower than those of using SodSulp and SodAcet. Thus higher amount of added NH_4^+ (during pH control) inhibited growth, sporulation of Btk through nitrogen catabolite repression [27-28]. Yezza et al. [13] also found a lower value of μ_{max} in case of AmmoAcet than SodSulp as pH control agents while using wastewater sludge as a raw material.

The amount of SO_4^{2-} used in case of SodSulp and AmmoSulp were the same (28.8 g) and the amount of CH_3COOH used in case of AmmoAcet and SodAcet was 73.75 and 67.85 g in 10 L working volume of bioreactor, respectively. This result revealed that the quantity of CH_3COO^- added into the medium was higher than that of SO_4^{2-}. It is well-known that CH_3COOH can serve as a carbon source for growth and synthesis of delta-endotoxin of Bt [20, 21]. Thus higher availability of easily assimilable carbon in the medium (in form of acetic acid) could be the reason for higher entomotoxicity at the end of fermentation (Table 7).

2.1.3.1.3. Alkaline protease (AP) and amylase activities in relation to different pH control agents

Alkaline protease activity

Bt produces multiple type of proteases which can contribute to the hydrolysis of protein components in the medium to amino acid that can be used for the growth and synthesis of delta-endotixin of Bt [19, 29-30]). Therefore, AP production profiles were recorded in this study. AP production started from 9h of fermentation irrespective of the type of pH control agents (Fig. 8b). At the peak (15 h) of protease activity profile, the AP activity was higher with NH_4OH compared to NaOH in combination with either acid (H_2SO_4 or CH_3COOH) as pH control agents. These results are in agreement with Zouari et al. [21] who found that ammonium sulfate did not improve bioinsecticide synthesis but increased proteolytic activity. Yezza et al. [13] also found that the protease activity was enhanced by 12 to 33% while replacing SodSulp with AmmoAcet as pH control agents when wastewater sludge and semi-synthetic soy were employed as growth media. Further, AP activity kept on increasing until the end of fermentation process and could be due to the lysis of completely sporulated cells and release of intracellular proteases into the medium [19, 25]. Thus, as indicated before, NH_4^+ may act as an inhibitor for growth and sporulation of Btk. However, NH_4^+ can play an important role to produce extracellular protease during growth of Btk in SIW medium. This may be one of the reasons for higher consumption of NH_4OH (AmmoAcet and AmmoSulp as pH control agents) than NaOH (SodSulp and SodAcet) during fermentation, as discussed before (Table 8).

Tableau 7. Maximum values of some parameters related to the fermentation process and the growth of Btk in SIW medium with different pH control agents

Parameters	pH control agents			
	SodSulp	*AmmoAcet*	*AmmoSulp*	*SodAcet*
Max. specific growth rate (μ max h^{-1})**	0.430	0.400	0.394	0.468
Max. total cell count (x 10^8CFU/ml) (time)	1.50 ± 0.35^a (12h)	5.70 ± 0.37^a (12h)	2.65 ± 0.18^c (9h)	3.80 ± 0.25^b (9h)
Cell count at 24h (x 10^8CFU/ml)***	3.20 ± 0.21^a	3.30 ± 0.23^a	2.50 ± 0.17^b	3.50 ± 0.22^a
Max. spore count at 24h (x 10^8 CFU/ml)***	1.35 ± 0.08^a	1.30 ± 0.08^a	1.00 ± 0.06^b	1.43 ± 0.09^a
Total cell count at 48h (x 10^8 CFU/ml)***	2.20 ± 0.14^a	1.90 ± 0.12^a	1.40 ± 0.09^b	2.10 ± 0.13^a
Spore count at 48h (x 10^8 CFU/ml)***	1.13 ± 0.07^a	0.80 ± 0.05^b	0.62 ± 0.04^c	1.20 ± 0.08^a
Max. sporulation at 48h (%)***	51.37 ± 0.20^b	42.10 ± 0.13^d	44.29 ± 0.13^c	57.13 ± 0.29^a
Delta-endotoxin concentration in fermented broth at 48h (mg/L)***	485.0 ± 24.2 ^{b}	450.5 ± 22.5 ^{bc}	432.8 ± 21.6 ^{c}	1043.0 ± 47.8 ^{a}
Tx of fermented broth at 48h (x 10^9 SBU/L) ***	16.70 ± 1.24 ^{ab}	14.30 ± 1.03 ^{bc}	13.10 ± 1.00 ^{c}	18.4 ± 1.38 ^{a}

SpTx ($x10^3$SBU/µg delta-endotoxin) in fermented broth	34.43	31.74	30.30	17.64
*Delta-endotoxin concentration in suspended pellet (mg/L)*****	4850 ± 242	4505 ± 225	4328 ± 216	10430 ± 478
Tx of suspended pellet (x 10^9 SBU/L) ***	21.20 ± 1.7 [b]	20.70 ± 1.6 [b]	19.40 ± 1.6 [b]	26.70 ± 2.1 [a]
SpTx ($x10^3$SBU/µg delta-endotoxin) in suspended pellet	4.40	4.60	4.50	2.56
SpTx-spore (SBU/1000 spore) in fermented broth at 48 h	147.8	178.8	211.3	153.3
*Spore concentration in suspended pellet ($x10^8$ CFU/ml)****	10.7 ± 0.8	7.6 ± 0.6	5.9 ± 0.4	11.4 ± 0.9
SpTx-spore (SBU/1000 spore) in suspended pellet	19.7	27.2	32.9	23.4

* Digits in parenthesis represent the time of fermentation at which maximum values of different parameters occurred.

** The presented values are the mean values obtained from two separately experiments conducted for each fermentation condition.

*** The values are the mean of three determination of two separately experiments conducted for each fermentation condition. The presented valued are the mean ± SD. Different letters (stand near each value) within the same row indicate the significant differences among these values determined by one-factor analysis of variance (Turkey HSD test, $p \leq 0.05$).

Amylase activity

Amylase production during the Btk fermentation is important because it can support the growth and synthesis of delta-endotoxin and other components of Btk through hydrolysis of residual starch present in SIW into glucose. Therefore, amylase production profiles were recorded in this study (Fig 8c). The amylase activity with SodSulp and SodAcet as pH control agents demonstrated diauxic type behaviour. However, in case of AmmoAcet and AmmoSulp as pH control agents, diauxic behaviour was less visible and amylase production started at 12 h of the fermentation process.

Tableau 8. Quantity of different pH control agents added during Btk fermentation process using SIW as a raw material in a 10 L working volume fermentor

pH control agents	Component	Volume (ml)*	Mole	Mole of H^+ or OH^-	Concentration of minerals/nutrients fortified (g/10L medium)
SodSulp	NaOH	25	0.10	0.10	$Na^+ = 2.30$
	H_2SO_4	100	0.30	0.60	$SO4^{2-} = 28.8$
AmmoAcet	NH_4OH	35	0.30	0.30	$NH_4^+ = 5.40$
	CH_3COOH	125	1.25	1.25	$CH_3COO^- = 73.75$
AmmoSulp	NH_4OH	30	0.25	0.25	$NH_4^+ = 4.50$
	H_2SO_4	100	0.30	0.60	$SO4^{2-} = 28.8$
SodAcet	NaOH	35	0.14	0.14	$Na^+ = 3.22$
	CH_3COOH	115	1.15	1.15	$CH_3COO^- = 67.85$

*The presented values are the mean values obtained from two separately experiments conducted for each fermentation condition.

It appears that due to the utilisation of RS (mostly glucose) present in SIW there was a delay in induction of amylase activity. This delay partly could also be due to Btk growth inhibition by NH_4OH in combination with either of the acids, as discussed before. The inhibition of amylase production by added NH_4OH was also evident from the fact that the highest amylase activity as well as higher production rate was observed in case of SodSulp as pH control agents (Fig. 8c). Further, the amylase activity in case of SodAcet was lower than SodSulp as pH control agent. This might be due to the fact that CH_3COOH could be used as a carbon source for growth of Btk and thus there was a lesser need to produce more amylase enzyme to hydrolyse available starch (into simple sugars like glucose) in the SIW.

2.1.3.2. Delta-endotoxin production, entomotoxicity of fermented broth and suspended pellet in relation to different pH control agents

2.1.3.2.1. Delta-endotoxin production

It is known that Bt delta-endotoxin is one of the main components for entomocidal activity of Bt biopesticides [19-21]. The change in nutrient components or fermentation process could affect the delta-endotoxin synthesis by Bt [20, 21], therefore, delta-endotoxin concentration was measured in all experiments using different pH control agents (Table 7). The delta-endotoxin concentration at the end of fermentation in different fermented broth using different pH control agents was different. Highest concentration of delta-endotoxin was obtained in case of SodAcet. CH_3COOH in combination with NaOH proved to be an excellent carbon source for delta-endotoxin synthesis. Moreover, it also demonstrated that use of NH_4OH (combined with either H_2SO_4 or CH_3COOH) as pH control agent caused a reduction in delta-endotoxin synthesis (Table 7).

2.1.3.2.2. Entomotoxicity of fermented broth

The variation of entomotoxicity (Tx) of fermented broth during fermentation using different pH control agents are depicted in Fig. 1d. In general, Tx increased with fermentation time until the end of the process irrespective of different pH control agents. Tx value with AmmoAcet was the lowest from 12 h to 36 h, however, increased rapidly between 36 – 48h and became slightly higher than the Tx value with AmmoSulp at this period. The profiles of Tx obtained using NaOH in combination with different acids (H_2SO_4 and CH_3COOH) to control pH were similar to those of NH_4OH in combination with H_2SO_4 or CH_3COOH (Fig. 8d). From 12 h to 36 h of fermentation process, SodSulp showed higher Tx than SodAcet, whereas from 36 h to 48h, the Tx value for SodAcet surpassed that of SodSulp. A higher Tx value in case of CH_3COOH could be due to the availability of additional carbon added in the form of CH_3COOH for growth, sporulation as well as for the spore maturation. Zouari and Jaoua [20] and Zouari et al. [21] also reported that supplementation of gruel-based medium with acetate enhanced the delta-endotoxin yield.

The Tx at the end of fermentation process was lower when CH_3COOH was used in combination with NH_4OH than in combination with NaOH. This difference was statistically significant (Table 7). As we discussed above, NH_4^+ showed inhibitory effect on growth and the sporulation process of Btk and consequently affected the entomotoxicity. Thus, unlike sludge, where AmmoAcet combination resulted in higher Tx value [13], the use of NH_4^+ as pH control agent in SIW negatively influenced the growth, sporulation, delta-endotioxin synthesis and Tx value. On the other hand, Na^+ has been reported to be an important mineral ion to generate entomotoxicity of Bt [31]. Therefore, combination of NaoH and CH_3COOH resulted in the highest entomotoxicity.

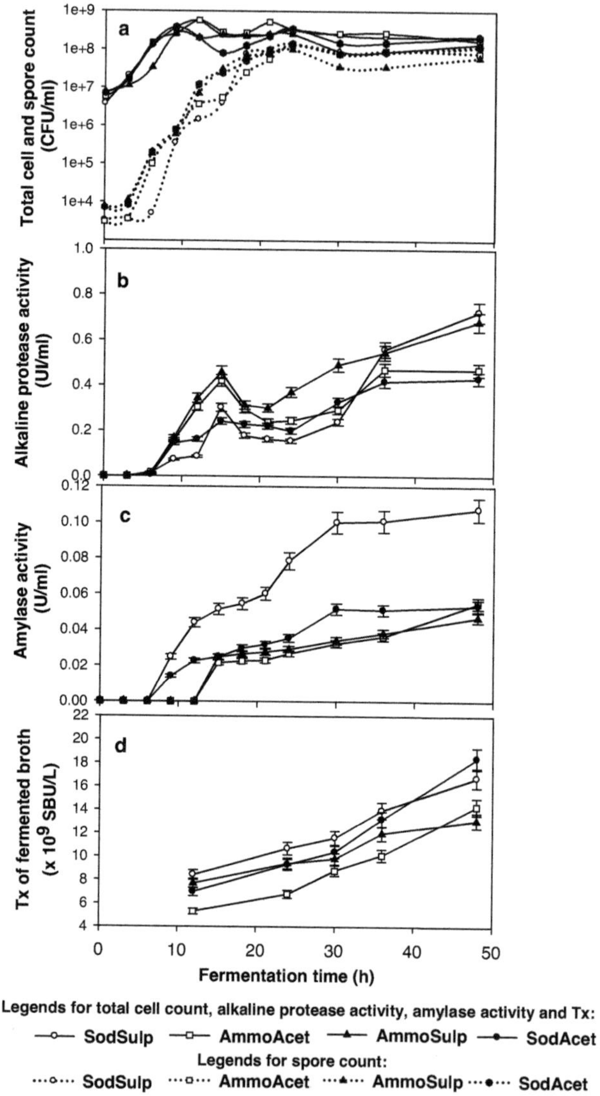

Figure 8. Profiles of (a) total cell count and spore count; (b) alkaline protease activity; (c) amylase activity and (d) entomotoxicity of fermented broth during Btkfermentation of SIW with different pH control agents

Yezza et al. [13] found entomotoxicity was increased using AmmoAcet as pH control agents during the fermentation of wastewater sludge from two different wastewater treatment plants located in Quebec (CUQS) and in Jonquiere (JQS) as compared to traditional pH control agents (SodSulp). However, in this research, the result was contrary. The possible reason is that the concentration of ammonia-nitrogen in case of SIW (26.7 mg/L in the medium) was higher than the concentration of ammonia-nitrogen in CUQS (15.4 mg/L in the medium) and JQS (12.0 mg/L in the medium) [13]. Thus, ammonia-nitrogen available in SIW + NH_4OH (supplied by using AmmoAcet or AmmoSulp) became higher than that of ammonia-nitrogen required for Btk and therefore became an inhibitory for Btk.

2.1.3.2.3 Entomotoxicity of suspended pellet (complex of spore and delta-endotoxin)

The entomotoxicity values of suspended pellets (10 times concentrated suspension of spore and delta-endotoxin complex) obtained after centrifugation of different fermented broths (different pH control agents) are presented in Table 7. The suspended pellets from each medium revealed significantly higher Tx values as compared to their respective fermented broths. The Tx value in case of SodAcet was the highest and statistically significantly different compared to three other cases (which were not statistically significantly different) (Table 7). The results also confirmed the role of delta-endotoxin and spore complex in increasing the Tx value as mentioned by other research [23]. Moreover, this research also demonstrated clearly that by using a simple method (supply carbon source as acetate) during fermentation, one could obtain a Bt based biopesticide with high potency.

2.1.3.3. Entomotoxicity versus delta-endotoxin and spore concentration

2.1.3.3.1 Entomotoxicity versus delta-endotoxin and spore concentration in fermented broth

Tx of fermented broth increased with delta-endotoxin concentration (Table 7). However, the increase in Tx value was not proportional to the delta-endotoxin concentration. For example, in case of SodAcet, delta-endotoxin concentration (1043.0 mg/L) was significantly higher than that of SodSulp (485.0 mg/L) but the Tx value was not proportionally higher.

Specific entomotoxicity (or SpTx) (SBU/μg delta-endotoxin) of four different fermented broths was calculated (Table 7). The values of SpTx in three cases (SodSulp, AmmoAcet and AmmoSulp) were not significantly different; however, it was nearly two times higher than SodAcet. Moreover, specific entomotoxicity versus spore (or SpTx-spore) (SBU/1000 spores) of four different fermented broth (at 48h) was also calculated (Table 7) and these values of SpTx-spore were different depending on fermented broth. Thus, it is difficult to predict Tx value based only on spore or delta-endotoxin concentration in fermented broth.

In fact, *Bacillus thringiensis* var. *kurstaki* HD-1 (Btk) used in this study is known to produce other agents (vegetative insecticidal proteins-Vip3A, immune inhibitor A and Zwittermicin A [7, 32-35]) that play synergistic role in biopesticidal activity. Moreover, this strain doesn't produce β-exotoxin – a toxin that is prohibited to be used as biopesticide [36]. Thus, there could be possible synergistic action between delta-endotoxin with spore and other compounds (Vip3A, immune inhibitor A and Zwittermicin A) possibly produced by Btk and thus affected the overall Tx of different fermented broths. These synergistic compounds might be produced in different concentration employing different pH control agents and that might contribute to different SpTx values. Also, the quality of delta-endotoxin produced by Bt in different fermented broths might be different [37] leading to different SpTx values.

2.1.3.3.2 Entomotoxicity versus delta-endotoxin and spore concentration in suspended pellet

SpTx (SBU/μg delta-endotoxin) of suspended pellet from 4 different fermented broths obtained using different pH control agents were presented in Table 2. Results showed that entomotoxicity of suspended pellet obtained by centrifuging the fermenteb broth using SodAcet as pH control agents was the highest but the SpTx value of suspended pellet of this fermented broth was the lowest as compared to other suspended pellets. However, the values of SpTx of three suspended pellets of fermented broths using SodSulp, AmmoAcet and AmmoSulp as pH control agents were not much different (Table 7). This fact is also supported by the research of Liu and Bajpai [37] who demonstrated that the changes in medium compositions could cause a change in specific activity of delta-endotoxin against the same target insect pest. Thus, it is possible that using the different pH control agents might cause the change in specific activity of delta-endotoxin against spruce budworm larvae used in this research.

In case of SpTx-spore (SBU/1000 spores) of suspended pellet from 4 different fermented broths (Table 7), these values of SpTx-spore were different depending on suspended pellet. Therefore, evaluation of potency of Bt based biopesticides could not be based on spore concentration. The results obtained in this research further confirmed that bioassay against the target insect pest is the only method that must be used to evaluate the biopesticidal activity of a Bt based biopesticide.

2.1.4. Conclusions

Based on the foregoing study employing different chemical agents to control pH during batch *Bacillus thuringiensis* fermentation of starch industry wastewater, the following conclusions could be drawn:

- NH_4OH showed a negative impact on growth, sporulation, delta-endotoxin synthesis and finally entomotoxicity of *Bacillus thuringiensis*, whereas, a positive effect on the alkaline protease production was observed.
- CH_3COOH provided additional carbon for growth, sporulation, delta-endotoxin synthesis and subsequently enhanced the entomotoxicity.
- Combination of NaOH and CH_3COOH as pH control agents during the fermentation process gave the highest delta-endotoxin concentration and final highest entomotoxicity of suspended pellet as well as fermented broth.
- Evaluation entomotoxicity of any Bt based biopesticides must be based on bioassay against the target insect pest.

2.1.5. Acknowledgements

The authors are sincerely thankful to the Natural Sciences and Engineering Research Council of Canada (Grant A4984, Canada Research Chair) for financial support. The views and opinions expressed in this article are strictly those of authors. Sincere thanks to Dr. Satinder K. Brar and Dr. Simon Barnabé for reading and providing suggestions to prepare the manuscripts.

2.1.6. References

1. Sikdar DP, Majumdar MK, Ajumdar SKM (1991) Effect of minerals on the production of the delta endotoxin by *Bacillus thuringiensis* subsp. *israelensis*. Biotechnol Lett 13: 511-514
2. Pearson D, Ward OP (1988) Effect of culture conditions on growth and sporulation of *Bacillus thuringiensis* subsp. *israelensis* and development of media for production of the protein crystal endotoxin. Biotechnol Lett 10: 451-456
3. Jin B, Huang LP, Lant P (2003) *Rhizopus arrhizus* - a producer for simultaneous saccharification and fermentation of starch waste materials to L(+)-lactic acid. Biotechnol Lett 25: 1983–1987

4. Jin B, van Leeuwen HJ, Patel B, Yu Q (1998) Utilisation of starch processing wastewater for production of microbial biomass protein and fungal α-amylase by *Aspergillus oryzae*. Bioresource Technol 66: 201-206

5. Huang LP, Jin B, Lant P, Zhou J (2003) Biotechnological production of lactic acid integrated with potato wastewater treatment by *Rhizopus arrhizus*. J Chem Technol Biotechnol 78: 899-906

6. Brar SK, Verma M, Tyagi RD, Valéro JR, Surampalli RY (2006) Efficient centrifugal recovery of *Bacillus thuringiensis* biopesticides from fermented wastewater and wastewater sludge. Water Res 40: 1310-1320

7. Brar SK, Verma M, Tyagi RD, Valéro JR (2006) Recent advances in downstream processing and formulations of *Bacillus thuringiensis* based biopesticides. Process Biochem 41: 323–342

8. Avignone Rossa C, Yantorno OM, Arcas JA, Ertola RJ (1990) Organic and inorganic nitrogen source ratio effects on *Bacillus thuringiensis* var. *israelensis* delta-endotoxin production. World J Microbiol Biotechnol 6: 27-31

9. Kraemer-Schafhalter A, Moser A (1996) Kinetic study of *Bacillus thuringiensis* in lab-scale batch process. Bioprocess Eng 14: 139-144

10. Harms RL, Martinez DR, Griego VM (1986) Isolation and characterization of coproporphyrin produced by four subspecies of *Bacillus thuringiensis*. Appl Environ Microbiol 51: 481-486

11. H.M. Nakata (1966) Role of acetate in sporogenesis of *Bacillus cereus*. J Bacteriol 91: 784-788

12. Jong JJ, Wu WT, Tzeng YM (1994) pH control for fed-batch culture of *Bacillus thuringiensis*. Biotechnol Techniques 8: 483-486

13. Yezza A, Tyagi RD, Valéro JR, Surampalli RY (2005) Influence of pH control agents on entomotoxicity potency of *Bacillus thuringiensis* using different raw material. World J Microbiol Biotechnol 21: 1549-1558

14. APHA, AWWA, WPCF (1998) Standard methods for examination of water and wastewaters, 20th ed., In Clesceri LS, Greenberg AE, Eaton AD (ed), American Public Health Association, Washington, DC

15. Avignone-Rossa C, Arcas J, Mignone C (1992) *Bacillus thuringiensis*, sporulation and δ-endotoxin production in oxygen limited and nonlimited cultures. World J Microbiol Biotechnol 8: 301–304

16. Aiba S, Humphrey AE, Millis NF (1973) Biochemical Engineering, Academic Press, 2nd ed., New York, 1973

17. Kunitz M (1947) Crystalline soybean trypsin inhibitor. J Gen Physiol 30: 291-310

18. Miller G (1959) Use of dinitrosalicylic acid reagent for determination of reducing sugar. Analytical Chem 31: 426-428

19. Zouari N, Jaoua S (1999) Production and characterization of metalloproteases synthesized concomitantly with delta-endotoxin by *Bacillus thuringiensis* subsp. *kurstaki* strain grown on gruel-based media. Enz Microb Technol 25: 364-371

20. Zouari N, Jaoua S (1999) The effect of complex carbon and nitrogen, salt, Tween-80 and acetate on delta-endotoxin production by a *Bacillus thuringiensis* subsp. *kurstaki*. J Ind Microbiol Biotechnol 23: 497-502

21. Zouari Z, Ben Sik Ali S, Jaoua S (2002) Production of delta-endotoxin by several *Bacillus thuringiensis* strains exhibiting various entomocidal activities towards lepidoptera and diptera in gruel and fish-meal media. Enz Microb Technol 31: 411-418

22. Bradford MM (1976) A rapid and sensitive for the quantitation of microgram quantitites of protein utilizing the principle of protein-dye binding. Analytical Biochem 72: 248-254

23. Dulmage HT, Boening OP, Rehnborg CS, Hansen GD (1971) A proposed standardized bioassay for formulations of *Bacillus thuringiensis* based on the international unit. J Invertebr Pathol 18: 240-245

24. Beegle CC (1990) Bioassay methods for quantification of *Bacillus thuringiensis* δ- endotoxin, analytical chemistry of *Bacillus thuringiensis*, pp 14-21. In Hickle LA, Fitch WL (ed), Analytical chemistry of *Bacillus thuringiensis*, American Chemical Society, Washington, DC
25. Bulla LAJr, Kramer KJ, Davidson LI (1977) Characterization of the entomocidal parasporal crystal of *Bacillus thuringiensis*. J Bacteriol 130: 375-383
26. Monteiro SM, Clemente JJ, Henriques AO, Gomes RJ, Carrondo MJ, Cunha AE (2005) A procedure for high-yield spore production by *Bacillus subtilis*. Biotechnol Progress 21: 1026-1031
27. Arcas J, Yantorno OM, Ertola RJ (1987) Effect of high concentration of nutrients on *Bacillus thuringiensis* cultures, Biotechnol Lett 9: 105-110
28. Fisher SH (1999) Regulation of nitrogen metabolism in *Bacillus subtilis*: Vive la difference. Mol Microbiol 32: 223-232
29. Donovan WP, Tan Y, Slaney AC (1997) Cloning of the *nprA* gene for neutral protease A of *Bacillus thuringiensis* and effect of in vivo deletion of *nprA* on insecticidal crystal protein. Appl Environ Microbiol 63: 2311–2317
30. Tyagi RD, Sikati Foko V, Barnabé S, Vidyarthi A, Valéro JR (2002) Simultaneous production of biopesticide and alkaline proteases by *Bacillus thuringiensis* using wastewater as a raw material. Water Sci Technol 46: 247-54
31. Drehval OA, Chervatiuk NV, Cherevach NV, Vinnikov AI (2003) Effect of mineral nutrition sources on the growth and toxin formation of the entomopathogenic bacteria *Bacillus thuringiensis*. Mikrobiol Z 65: 14-20 (English Translation)
32. Donovan WP, Donovan JC, Engleman JT (2001) Gene knockout demonstrates hat *vip3A* contributes to the pathogenesis of *Bacillus thuringiensis* toward *Agrotis ipsilon* and *Spodoptera frugiperda*. J Invertebr Pathol 78: 45–51

33. Guttmann DM, Ellar DJ (2000) Phenotypic and genotypic comparisons of 23 strains from the *Bacillus cereus* complex for a selection of known and putative *B. thuringiensis* virulence factors. FEMS Microbiol Lett 188: 7-13

34. Manker DC, Lidster WD, Starnes RL, MacIntosh SC (1994) Potentiator of *Bacillus* pesticidal activity. Patent Coop Treaty WO94/09630

35. Stabb EV, Jacobsen LM, Handelsman J (1994) Zwittermicin A-producing strains of *Bacillus cereus* from diverse soils. Appl Environ Microbiol 60: 4404-4412

36. Levinson BL, Kasyan KJ, Chiu SS, Currier TC, Gonzalez JMJr (1990) Identification of β-exotoxin production, plasmids encoding β-Exotoxin, and a new exotoxin in *Bacillus thuringiensis* by using high-performance liquid chromatography. J Bacteriol 172: 3172-3179

37. Liu WM, Bajpai RK (1995) A Modified Growth Medium for *Bacillus thuringiensis*. Biotechnol Prog 11: 589-591

PARTIE 2.2. Les eaux usées d'industrie d'amidon (SIW) pour la production de biopesticides - Effets des concentrations de matières solides

Résumé

Diverses concentrations de solides totaux (TS), allant de 15 à 66 g/L contenus dans les eaux usées d'industrie d'amidon (SIW) ont été testées comme matières premières pour la production de biopesticide à base de *Bacillus thuringiensis* (Btk HD-1) par fermentation en erlenmeyers et en bioréacteurs de 15 L. La fermentation en erlenmeyer a donné la plus forte concentration en delta-endotoxines pour un TS de 30 g/L et à 2,5% de volume de pré-culture (v/v). Les expériences en fermenteurs en utilisant les SIW (30 g/L en TS) ont été réalisées dans des conditions contrôlées de température, de pH et d'oxygène dissous et ont conduit à obtenir des concentrations plus élevées en spores, en delta-endotoxines et en enzymes (protéases et amylases), comparativement à ce qui est atteint avec les SIW (15 g/L en TS) comme contrôle. La valeur maximale de la Tx (déterminée par bioessais contre des larves de la tordeuse des bourgeons de l'épinette) des bouillons fermentés de Btk HD-1 a été observée avec des SIW contenant 30 g/L de TS et suivi dans l'ordre: SIW 30 g/L TS (17,8 x 10^9 SBU/L) > SIW 15 g/L TS (SBU 15,3 x 10^9 SBU/ L) > milieu synthétique (11,7 x10^9 SBU/L). Le complexe de spores et de delta-endotoxines obtenu après centrifugation suivie par une remise en suspension (avec le surnageant), en un dixième du volume initial du bouillon fermenté des SIW ayant 30 g/L en TS, présente le plus grand potentiel d'application (pour la protection des forêts contre la tordeuse des bourgeons de l'épinette) que les autres milieux et ce, en terme d' entomotoxicité.

Starch industry wastewater for production of biopesticides – Ramifications of solids concentrations

Abstract

Total solids (TS) concentrations ranging from 15 to 66 g/L of starch industry wastewater (SIW) were tested as raw material for the production of *Bacillus thuringiensis* var. *kurstaki* HD-1 (Btk) biopesticide in shake flasks and 15L bench scale fermentor. Shake flask studies revealed higher delta-endotoxin concentration of Btk at 30 g/L TS concentration and 2.5% (v/v) volume of pre-culture. The fermentor experiments conducted using SIW 30 g/L TS concentration under controlled conditions of temperature, pH and dissolved oxygen showed higher spore count, enzyme production (protease and amylase), delta-endotoxin concentration as compared to those of SIW 15 g/L TS concentration. The entomotoxicity, at the end of fermentation, with SIW 30 g/L solids concentration (17.8×10^9 SBU/L, measured against spruce budworm) was considerably higher as compared to entomotoxicity at 15 g/L solids concentration (15.3×10^9 SBU/L) and semi-synthetic medium (11.7×10^9 SBU/L). The pellet comprising spores and delta-endotoxin complex obtained after centrifugation and followed by re-suspension (in supernatant) in one tenth of the original volume of SIW 30 g/L solids concentration media registered highest potential for application (to protect forests against spruce budworm) than other media in term of entomotoxicity.

Key words: *Bacillus thuringiensis* var. *kurstaki* HD-1; delta-endotoxin; entomotoxicity; solids concentration; starch industry wastewater

2.2.1. Introduction

Growth and synthesis of bacterial metabolites need optimal nutrients, especially, in the case of production of *Bacillus thuringiensis* (Bt) based biopesticides [1]. The semi-synthetic media for Bt biopesticides production selected based on the usage of the optimal nutrients has been investigated. For example, media based on glucose concentration [1-3] or based on optimal ratio of carbon: nitrogen at different initial total carbon [4] has been reported. Maximum protein and delta-endotoxin yields were obtained in a semi-synthetic medium that contained glucose concentration of 6 to 8 g/L [2]. Arcas et al. (1987) evaluated that glucose-yeast extract medium increased spore counts from 1.08 x 10^{12} spores/L to 7.36 x 10^{12} spores/L and toxin level from 1.05 mg/ml to 6.85 mg/ml, when the concentration of glucose was increased from 8 to 56 (g/L), with the corresponding increase in the rest of medium components. Higher concentration of nutrients normally inhibits either spore count or toxin production [3].

In case complex medium such as wastewater sludge or other waste materials for Bt biopesticides production, the nutrients required for Bt growth are present in solids. Thus, an increase the solids concentration in the medium enhanced carbon and other nutrients and finally hence the entomotoxicity (Tx) [5-6]. It was found that the sludge solids concentration significantly influenced growth and Tx yield. The optimum total solids concentration was 26 g/L, which resulted in an improved potency of 12.9 x 10^9 IU/L, cell and spore concentrations of 5.0x10^9 and 4.8x10^9 CFU/ml, respectively, and a sporulation rate of 96% [5].

Utilization of starch industry wastewater (SIW) for Bt based biopesticides production offered several advantages as compared to other semi-synthetic medium or sludge such as raw material is free of contaminants and prospects of high Tx. In fact, SIW can be used directly without any pre-treatment

[7,10]. However, in those previous researches the impact of SIW solids concentration on Bt growth, enzyme production, delta-endotoxin production and Tx has not yet been investigated. SIW contains approximately 15 g/L total solids concentration and has been employed as such to produce Bt based biopesticides. Hence, the principal objective of this research article was to explore the ramification of solids concentration on Bt based biopesticides production using starch industry wastewater as raw material. Cell and spore count, enzyme production (protease and amylase), delta-endotoxin concentration, Tx of fermented broth as well as Tx of suspended pellet (obtained from fermented broth) were used as analytical parameters.

2.2.2. Material and methods

2.2.2.1. Bacterial strain and biopesticide production medium

Bacillus thuringiensis var. *kurstaki* HD-1 (ATCC 33679) (Btk) was used in this study. The Btk was subcultured and streaked on tryptic soya agar (TSA) plates, incubated for 48 hour at 30 ± 1°C and then preserved at 4.0 ± 1 °C for future use [6].

During the course of this study, starch industry wastewater (SIW) from ADM-Ogilvie (Candiac, Québec, Canada) was used as a raw material for Btk growth. Total solids (TS), volatile solids (VS), suspended solids (SS) and volatile suspended solids (VSS), pH, total carbon, total Kjeldahl nitrogen (TKN), ammonia nitrogen ($N{-}NH4^+$), total phosphorus and metals concentration (Al, Ca, Cd, Cr, Cu, Fe, Mg, Mn, Na, Ni, Pb and Zn) were determined according to Standard Methods [11]. Physico-chemical characteristics of SIW are presented in Table 9.

Tableau 9. Characteristics of starch industry wastewater (SIW)

Parameter	Starch industry wastewater
Total solids [TS] (g/L)	15.0 ± 0.7
Volatile solids [VS] (g/L)	10.6 ± 0.5
Suspended solids [SS] (g/L)	5.2 ± 0.2
Volatile suspended solids [VSS] (g/L)	5.0 ± 0.2
pH	3.6 ± 0.2
	Concentration (g/kg TS)
Ct	397 ± 7
N_t	28 ± 1
$N\text{-}NH_3$	0.70 ± 0.01
P_t	7.0 ± 0.2
$P\text{-}PO_4^{3-}$	4.00 ± 0.02
Ca	4.00 ± 0.06
K	11.0 ± 0.2
Mg	3.00 ± 0.03
Na	84.0 ± 0.3
S	5.0 ± 0.1
	Concentration (mg/kg TS)
Al	193.8 ± 6.3
Cd	0.13 ± 0.02
Cr	5.53 ± 0.04
Cu	22.20 ± 0.26
Fe	332.5 ± 14.1
Mn	40.70 ± 0.26
Pb	1.06 ± 0.02
Zn	67.85 ± 0.63
Ni	2.48 ± 0.02

Different TS concentrations of SIW (15; 23; 30; 35; 43; 48; 55; 61 and 66 g/L) were obtained by sedimentation process of original SIW (15 g/L TS). They were used as raw materials for investigating the optimal TS for Bt biopesticide production. SS and Dissolved Solids (DS) of different TS concentrations are presented in Table 10.

Tableau 10. Variation of suspended solids (SS), dissolved solids (DS) and DS/SS ratio at different concentrations of total solids

TS (g/L)	15.0	23.0	30.0	35.0	43.0	48.0	55.0	61.0	66.0
SS (g/L)	5.20	11.27	19.8	24.10	32.40	37.40	42.0	47.28	52.20
DS (g/L)	9.85	11.73	10.20	10.90	10.60	10.60	13.0	13.72	13.80
DS/SS	1.91	1.0	0.52	0.45	0.33	0.28	0.31	0.29	0.27

Synthetic medium (semi-synthetic medium) based on soybean meal was also used as control medium. It consist of (g/L): soybean meal (15.0); glucose (5.0); starch (5.0), K_2HPO_4 (1.0); KH_2PO_4 (1.0); $MgSO_4.7H_2O$ (0.3); $FeSO_4.7H_2O$ (0.02); $ZnSO_4.7H_2O$ (0.02); $CaCO_3$ (1.0).

2.2.2.2. Inoculum preparation

All the media used for inoculum preparation were adjusted to pH 7 before autoclaving. A loopful of Btk from TSA plate was used to inoculate a 500-ml Erlenmeyer flask containing 100 ml of sterilised tryptic soya broth (TSB) medium. The flask was incubated on an incubator- shaker at 220 revolutions per min (rpm) and 30 ± 1 °C for 8–12 h. A 2% (v/v) inoculum from this flask (first stage) was then used to inoculate 500-ml Erlenmeyer flasks containing

100 ml of sterilised SIW at different TS concentrations (15; 23; 30; 35; 43; 48; 55; 61 and 66 g/L). The flasks were incubated for 8-12 h and the actively growing cells from these flasks were used as inoculum (second stage or pre-culture) for the production of Btk based biopesticide in shake flasks or in fermentor [10]. For investigating the effect of different inoculum volumes on the growth and delta-endotoxin synthesis, 2% (v/v) of inoculum (first stage) was used to inoculate 500-ml Erlenmeyer flasks containing 100 ml of sterilised SIW at optimal TS concentration. The flask was incubated for 8-12h and the actively growing cells from these flasks were used as inoculum (second stage or pre-culture) for the production of Btk based biopesticide in shake flasks or in fermentor.

2.2.2.3. Shake flask fermentation

2.2.2.3.1. Effect of solids concentration

Erlenmeyer flasks containing 100 ml of sterilised SIW at different TS concentration (15; 23; 30; 35; 43; 48; 55; 61 and 66 g/L) and inoculated with 2% (v/v) of pre-culture were prepared as above. The flasks were incubated in a shaker-incubator (New Brunswick) for 48 h at 30 ± 1 °C, 220 rpm. Samples were collected at the end of fermentation process to determine the total cell count (TC), spore count (SC), and delta-endotoxin concentration as described in the following sections. The values presented were the average results (± Standard Deviation (SD)) of triplicates of two independent experiments. From these results, the optimum solids concentration that resulted in the highest delta-endotoxin concentration was selected to investigate the effect of optimal inoculum volume on the Btk growth and delta-endotoxin synthesis.

2.2.2.3.2. Effect of inoculum volume

Different inoculum volumes of pre-culture (2.0; 2.5; 3.0; 3.5; 4.0; v/v) as prepared above were used for inoculating into the Erlenmayer flasks containing 100 ml of sterilized SIW at optimal TS concentration. The flasks were incubated in an incubator- shaker (New Brunswick) for 48 h at 30 ± 1

°C, 220 rpm. Samples were collected at the end of fermentation process to determine the TC, SC, and delta-endotoxin concentration as described in the following sections. The values presented were the average results (± SD) of triplicates of two independent experiments.

2.2.2.4. Fermentation procedure in 15-L computer controlled bioreactor

To evaluate and compared the overall effects of Btk based biopesticides production, fermentations on bioreactor were conducted using three media: synthetic medium, SIW at 15 g/L TS and SIW at 30 g/L (optimize obtained in shake flask)). Fermentation was carried out in a stirred tank 15 L fermentor (working volume: 10 L, Biogénie, Que., Canada) equipped with accessories and programmable logic control (PLC) system for dissolved oxygen (DO), pH, anti-foam, impeller speed, aeration rate and temperature. The software (iFix 3.5, Intellution, USA) allowed automatic set-point control and integration of all parameters via PLC. The fermentor was then inoculated with 2% v/v inoculums (optimal in cases of synthetic medium and SIW at 15 g/L TS) [10] and 2.5% v/v inoculums (in case of SIW at 30 g/L TS) aseptically with preculture of Btk in exponential phase (8-12 h). In order to keep the DO above 25% saturation, air flow rate and agitation rate were varied between 2-4 LPM and 250–500 rpm, respectively. This ensured the critical DO level for Bt above 25 % [12]. The temperature was maintained at 30 ± 1 °C by circulating water through the fermentor jacket. Fermentation pH was controlled automatically at 7 ± 0.1 through computer-controlled peristaltic pumps by addition of pH control agents: NaOH 4 M/H_2SO_4 3 M (sodium hydroxide 4 M/sulphuric acid 3 M). Dissolved oxygen and pH were continuously monitored by means of a polarographic dissolved oxygen probe and of a pH sensor (Mettler-Toledo, USA), respectively. Samples were collected periodically to monitor the changes in TC, SC, delta-endotoxin concentration and Tx as described in the following sections. A part of each sample (20 ml) was centrifuged at 8000 rpm in 20 min and 4°C. The supernatant thus obtained was used to determine alkaline protease activity and amylase activity.

2.2.2.5. Analysis parameters

2.5.1. Determination of volumetric oxygen transfer coefficient (K_La), Oxygen Transfer Rate (OTR) and Oxygen Uptake Rate (OUR)

The volumetric oxygen transfer coefficient (K_La) measurement was based on the dynamic method [13-14]. This technique consisted in interrupting the air input. Afterwards, the aeration was re-established. The decrease (aeration off) and the increase (aeration on) in DO concentration were recorded. K_La was determined from the mass balance on DO just after each sampling of the fermentation broth. During batch fermentation, the mass balance for the DO concentrations could be written as follows: $dC_L/dt = OTR - OUR$

OTR: represents the oxygen transfer rate from the gas phase to the liquid phase:

$OTR = K_La(C^*-C_L)$

OUR: represents the oxygen uptake rate:

$OUR = Q_{O2}X$

K_La: volumetric oxygen transfer coefficient

C^*: oxygen concentration at saturation

C_L: Dissolved oxygen concentration in the medium

Q_{O2} : specific oxygen uptake rate

X: cell concentration

Oxygen concentration in the fermentation broth was converted from % air saturation to mmol O_2/l as follows: the DO electrode was calibrated in the medium at 30 ± 1 °C and then transferred to air-saturated distilled water at known temperature and ambient pressure. This reading was used with the known saturation concentration of oxygen at saturation in distilled water

(0.07559 mmol/l) (100%), to estimate the saturation concentration of oxygen at saturation in the cultivation media at 30 °C.

2.2.2.5.2. Alkaline protease activity (PA) and amylase activity assay (AA)

PA was determined according to modified Kunitz method [15]. One proteolytic activity unit was defined as the amount of enzyme preparation required to liberate 1 μmol (181 μg) tyrosine from casein per minute in pH 8.2 buffer at 37 ± 1 °C. The values presented were the average results of three determinations of two separately experiments ± Standard Deviation (SD). AA was measured according to the method of Miller [16-17]. One enzyme unit was defined as the amount of enzyme required to liberate 1 mg of glucose/min under the experimental conditions. The values presented were the average results (± SD) of triplicates of two independent experiments.

2.2.2.5.3. Estimation of delta-endotoxin concentration produced during the fermentation

Delta-endotoxin concentration was determined based on the solubilization of insecticidal crystal proteins in alkaline condition [17-18]: 1 ml of samples collected during fermentation was centrifuged at 10 000 g for 10 min. at 4± 1 °C. The pellet containing a mixture of spores, insecticidal crystal proteins, cell debris and residual suspended solids was used to estimate the concentration of alkaline solubilised insecticidal crystal proteins (delta-endotoxin): These pellets were washed three times with 1 ml of 0.14 M NaCI - 0.0 1 % Triton X- 100 solution. The washing helped in eliminating the soluble proteins and proteases which might have adhered on the pellets and could affect the integrity of the crystal protein. The insecticidal crystal proteins in the pellet were solubilized with 0.05N NaOH (pH 12.5) for three hours at 30°C with stirring. The suspension was centrifuged at 10 000 g for 10 min. at 4 ± 1 °C and the pellet, containing spores, cell debris and residual suspended solids was discarded. The supernatant, containing the alkaline solubilised insecticidal crystal proteins was used for the determination of delta-endotoxin concentration by Bradford method using bovine serum albumin as standard protein [19]. The values presented were the average results (± SD) of triplicates of two independent experiments.

2.2.2.5.4. Estimation of TC and SC

The TC and SC were performed by counting colonies grown on TSA medium [10]. For all counts, the average of at least triplicate plates was used for each tested dilution. For enumeration, 30–300 colonies were enumerated per plate. The results were expressed as colony forming units per mL (CFU/ml).

2.2.2.5.5. Preparation of pellet for bioassay

At the end of fermentation, fermented broth of three media were adjusted to pH 4.5 and centrifuged at 9000 g for 30 min [8]. The individual pellets (spore and delta-endotoxin complex) so obtained were re-suspended with supernatant resulting in one-tenth volume of original fermented broth. The suspended pellets were used to determine Tx through bioassay.

2.2.2.5.6. Bioassay technique

The potential Tx of Btk was estimated by bioassay against third instar larvae of spruce budworm (Lepidoptera : *Choristoreuna fumiferena*) following the diet incorporation method of Dulmage et al. (1971) and Beegle (1990) [21-22]. The commercial preparation 76B FORAY (Abbott Laboratory, Chicago, IL) was used as a reference standard to analyse the Tx. The detail procedure of bioassay technique was presented in Yezza et al. (2006) [10]. The Tx was evaluated by comparing the mortality percentages of dilutions of the samples (suspended pellet as well as fermented broth during the fermentation of three media) with same dilutions of the standard. In our research, Tx was expressed as relative spruce budworm units (SBU/L). On comparison of Tx of Bt-fermented samples, it was found that SBU reported in our study was 20–25% higher than IU [10]. The presented values are the mean of three determinations of two independent experiments ± SD.

2.2.3. Results and discussion

2.2.3.1. Shake flask fermentation

3.1.1. Optimal solids concentration

The fermentation experiment at various TS concentrations (15, 23, 30, 35, 43, 48, 55, 61, 66 g/L) of SIW was conducted in shake flasks using inoculum volume of pre-culture at 2% (v/v) and results of TC, SC and delta-endotoxin at the end of fermentation are presented in Figure 9a. Results showed that increase in TS concentrations of SIW (from 15 to 43 g/L) caused an increase in the TC, however, further increase in TS concentrations (from 48 to 66 g/L), the TC considerably decreased. The SC increased in the range of TS concentrations from 15 to 35 g/L and decreased slightly at TS concentrations > 35 g/L. The delta-endotoxin concentration of fermented broth increased with TS concentration and attained the highest value at 30 g/L. The delta-endotoxin production at 35 g/L was less than that of compared to at 30 g/LTS, however, it was statistically insignificant ($p \leq 0.05$) (Figure 9a). Further increase in TS concentration resulted in considerable decrease of the delta-endotoxin concentration.

Increase in viscosity, and relatively higher degree of heterogeneity at higher TS concentration may cause limitation of oxygen transfer, specifically in shake flask leading to inhibition of Btk growth, spore formation and finally to the Tx [5-6]. Increase of osmotic pressure across the bacterial membrane due to increased solids concentration could also contribute towards inhibition of growth, sporulation and the delta-endotoxin production.

It is interesting to note that at different TS concentrations of SIW, DS concentration was not significantly different (Table 10); instead they were in the range of 9.90 - 13.80 g/L.

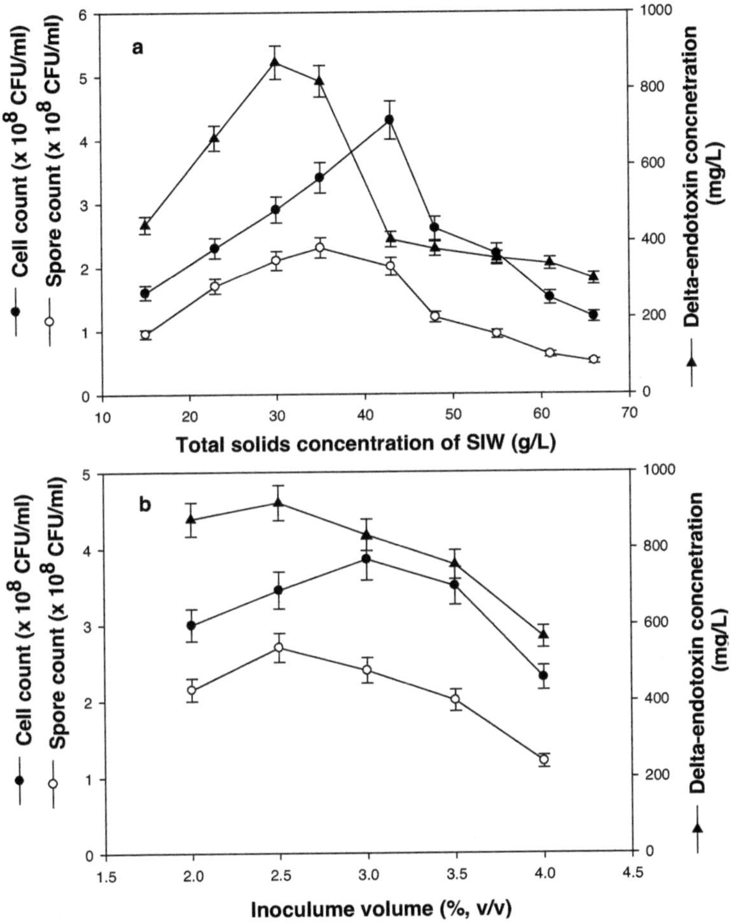

Figure 9 (a) Effects of different solids concentration (g/L) and (b) effects of different inocolum volumes of pre-culture (v/v) on the growth and delta-endotoxin concentration of Btk

The variation in SS concentration caused the changes in ratio of DS/SS between 0.27-1.91 (Table 10). The DS fraction of SIW might contain soluble components such as soluble protein, reducing sugars, minerals, etc and the SS fraction of SIW might contain insoluble component such as residual starch, complex proteins, fibres, etc. It is possible Btk during fermentation preferentially utilizes the DS fraction for growth and product production. Once DS is depleted or its concentration becomes low, bacteria will utilize the insoluble components or complex components present in the SS portion of the media. As mentioned above, TS in the range of 15 g/L to 35 g/L was more suitable for Btk growth and synthesis of delta-endotoxin. This range was corresponded to the ratios of DS/SS in the range of 0.45-1.91 (Table 10). At higher TS or lower ratio of DS/SS, the media became less effective for Btk growth and delta-endotoxin synthesis. This fact should be taken account when using SIW as raw material for fermentation of Btk as well as for other microorganisms (if have).

2.2.3.1.2. Optimal inoculum volume

Different inoculum volumes (2, 2.5, 3, 3.5, and 4.0 v/v) of pre-culture were used to inoculate the optimal solids concentration of SIW of 30 g/L for Bt based-biopesticides production during 48h. Data of TC, SC and delta-endotoxin concentration at the end of fermentation (48h) are presented in Figure 1b. Results showed that the highest delta-endotoxin concentration was obtained at inoculum volume of pre-culture of 2.5% (v/v). At higher inoculum volumes (3 to 4%, v/v), the spore and delta-endotoxin concentrations decreased. In fact, high initial cell concentration may result in a rapid consumption of oxygen and other nutrients [5-6]. Consequently, the limitation of dissolved oxygen resulted in lower cell and spore formation as well as delta-endotoxin synthesis. Moreover, oxygen is one of the most important factors for the growth, sporulation and delta-endotoxin synthesis by Bt [12].

2.2.3.2. Fermentation in bioreactor

The experiments in shake flasks showed that for the Btk biopesticides production (maximum delta-endotoxin synthesis) the optimal TS concentration and inoculum volume were 30 g/L and 2.5% (v/v), respectively. However, it is known that shake flask fermentation has major drawback of possible limitation in oxygen transfer. Therefore, experiments in environmentally controlled 15L-bioreactor were conducted to confirm the results at optimal TS concentration (30 g/L) and inoculums volume (2.5% v/v). Fermentation runs were also conducted employing synthetic medium and SIW (15 g/L TS) as raw materials to compare the overall results.

2.2.3.2.1. Fermentation parameters

Profiles of fermentation parameters, namely, dissolved oxygen concentration (DO, % saturation), agitation (revolution per minute - RPM), aeration (litre per minute- LPM), pH and temperature, volumetric oxygen transfer coefficient (K_La), oxygen uptake rate (OUR), oxygen transfer rate (OTR) for three different media (synthetic medium, SIW 15 g/L and SIW 30 g/L) are presented in Figure 10 and the maximum values of these parameters are presented in Table 11.

During the fermentation, agitation (250-500 RPM) and aeration (2-4 LPM) were controlled to maintain the DO above critical point (> 25 %) [12]. In the beginning of fermentation from 9 to 12h (depending on the medium), DO decreased followed by an increase until the end of the fermentation. The decrease in DO coincided with the exponential Btk growth phase which necessitated higher oxygen uptake rate (OUR). The increase of DO and decrease in OUR (9-12 h onwards) was due to the fact that Btk entered the stationary growth phase (synthesis of spore and delta-endotoxin) with lower oxygen demand.

The maximum values of K_La observed for different media were in the following order: 119 h^{-1} (SIW 30 g/L) > 84 h^{-1} (SIW 15 g/L) > 76.4 h^{-1} (synthetic medium) (Table 10). In fact, k_La values depend on factors such as aeration and agitation degree, rheological properties of culture broth and the presence (or concentration) of antifoam [14]. Figure 2 shows that agitation speed in case of synthetic medium and SIW 15 g/L were almost same, however, even in spite of higher aeration rate in synthetic medium than that of SIW (TS - 15 g/L), the maximum value of k_La was lower in synthetic medium. The possible reason for lower value of K_La in synthetic media could be attributed to different rheological characteristics of two media as a result of higher total solids concentration in synthetic medium (28.34 g/L) than that in SIW (15 g/L TS). Nevertheless, higher maximum value of k_La during fermentation of concentrated SIW (TS -30 g/L) was due to use of higher agitation and aeration rates than other cases (Figure 10).

2.2.3.2.2. Growth of Btk during fermentation

Profiles of total cell count (TC) and spore count (SC) during growth of Btk in different media are depicted in Figure 11a and the maximum values of growth parameters (maximum cell and spore counts, maximum specific growth rate, maximum % sporulation) are illustrated in Table 11. It was found that using optimal TS concentration (30 g/L) of SIW and optimal inoculum volume (2.5% v/v) caused considerable increase in growth of Btk as compared to non-concentrated SIW (15 g/L TS). Maxium TC observed in SIW 30 g/L TS was higher than TC in synthetic medium whereas maximum SC in SIW 30 g/L was almost similar as SC in synthetic medium (Table 11). However, at the end of fermentation (48h) TC and SC values in two media were almost similar.

Tableau 11. Maximum values of fermentation process parameters in different media

Parameters	Synthetic medium	SIW 15 g/L TS	SIW 30 g/L TS
Max. k_La *(h^{-1}) ***	76.4 h^{-1} (12h)*	84 h^{-1} (12h)	119 h^{-1} (15h)
*Max OTR (mmolO_2/L.h) ***	2.35	2.10	2.80
*Max. OUR (mmolO_2/L.h) ***	1.47	1.22	2.00
*Max. specific growth rate (μ max h^{-1}) ***	0.38	0.43	0.46
*Max. total cell count (x 10^8 CFU/ml) ****	6.0 ± 0.42^b (15h)	5.60 ± 0.36^b (12h)	8.30 ± 0.58^a (18h)
*Max spore count (x 10^8 CFU/ml) ****	3.4 ± 0.23^a (48h)	1.50 ± 0.10^b (24h)	3.0 ± 0.21^a (36h)
*Total cell count at 48h (x 10^8 CFU/ml) ****	3.8 ± 0.26^a	2.4 ± 0.17^b	3.5 ± 0.24^a
*Spore count at 48h (x 10^8 CFU/ml) ****	3.4 ± 0.24^a	1.50 ± 0.10^b	3.0 ± 0.21^a
*Max. sporulation at 48h (%)****	89.5 ± 6.3^a	60.0 ± 4.20^b	85.7 ± 5.9^a

*Max amylase activity (mU/ml)****	227 ± 15.4 [a] (18h)	100 ± 7.0 [c] (30h)	180 ± 10.8 [b] (15h)
*Max protease activity (UI/ml)****	1.55 ± 0.12 [a] (15h)	0.39 ± 0.03 [c] (15h)	0.80 ± 0.06 [a] (12h)
*Max delta-endotoxin concentration (mg/L) at 48h****	1020 ± 51 [b]	459.5 ± 22.9 [c]	1112.1 ± 55.6 [a]

* Digits in parenthesis represent the time of fermentation at which maximum values of different parameters occurred.

** The presented values are the mean values obtained from two separate experiments conducted for each fermentation condition.

*** The values are the mean of three determinations of two separate experiments conducted for each fermentation condition. The presented values are the mean ± SD. Different letters (stand near each value) within the same row indicate the significant differences among these values determined by one-factor analysis of variance (Turkey HSD test, $p \leq 0.05$).

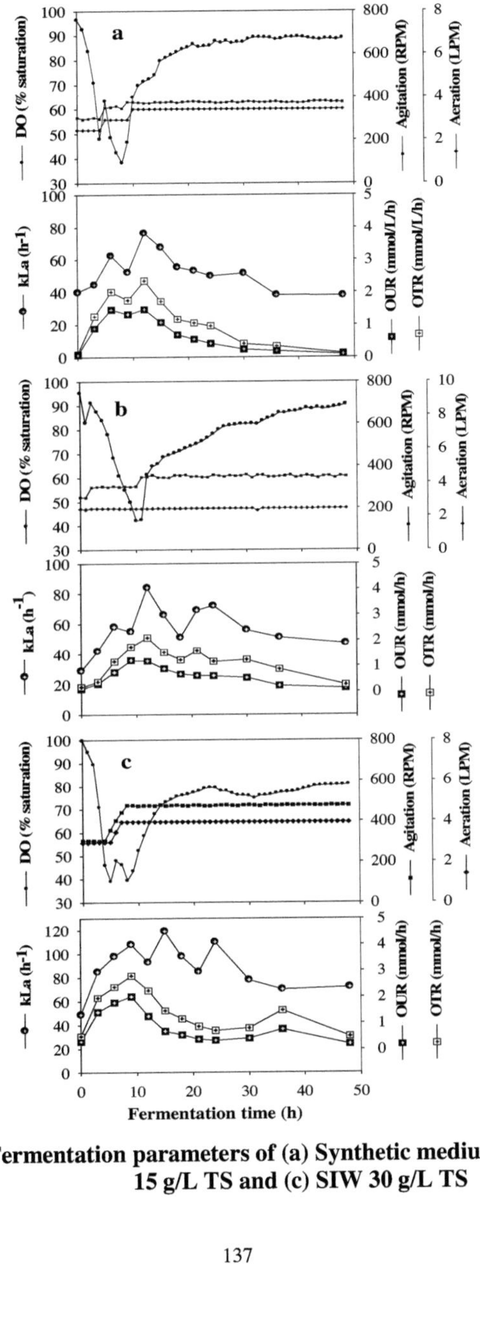

Figure 10. Fermentation parameters of (a) Synthetic medium; (b) SIW 15 g/L TS and (c) SIW 30 g/L TS

2.2.3.2.3. Enzyme production of Btk during fermentation

Increase in TS concentration of SIW from 15 g/L to 30 g/L caused an increase in amount of available nutrients as well as an increase in protein and starch concentration which induced synthesis of hydrolysing enzymes (protease and amylase) to convert these substrates into easily consumable nutrients (amino acids and glucose) for Btk growth and synthesis of spores and delta-endotoxin. Therefore, protease activity (PA) and amylase activity (AA) during the fermentation of three media were determined and their profiles as well as their maximum values were presented in Figure 11b and Table 11, respectively. AA and PA were highest when synthetic medium was employed. Moreover, maximum AA and PA activities increased with increase in TS concentration in SIW (Table 11). An increase in TS concentration in SIW increased the complex carbohydrates (starch) and proteins which lead to synthesis of higher PA and AA by Btk.

2.2.3.2.4. Delta-endotoxin production

In many studies, delta-endotoxin concentration produced by Bt in different media was normally determined at the end of fermentation [17-18]. This was based on the hypothesis that at the end of sporulation process, the sporulated cells (containing delta-endotoxin and spore and other intracellular components) will be lysed releasing delta-endotoxin, spore and other intracellular components into the medium [23-24]. However, we observed that sporulation occurred throughout the growth of Btk (Fig. 11a), therefore, in this research, variation of delta-endotoxin concentration produced was determined with the fermentation time in 3 different media and the various profiles of toxin thus obtained are presented in Figure 12. The delta-endotoxin concentration produced during Btk fermentation of SIW at 30 g/L TS was considerably higher than that in the synthetic media or SIW at 15 g/L TS. In three media, delta-endotoxin was present at the beginning of fermentation (low concentration) and increased during fermentation until the end of the process. From 6h to 24h, the increase in toxin concentration was considerably faster; however, during the period of 24h to 48h, the increase was relatively slow. A rapid increase of delta-endotoxin (from 6h to 24h) related to a rapid spore formation during this period (Figure 11a). From 24h to 48h, the profiles of delta-endotoxin concentration were different for (1) in SIW at 15 g/L TS, spore concentration stopped increasing (Figure 3a) which in turn arrested the increase of delta-endotoxin (Fig. 12); (2) in the case of synthetic medium and SIW at 30 g/L TS, there was slight increase in spore formation until 48h and 36h, respectively (Figure 11a and Table 11) with concomitant slow increase in delta-endotoxin concentration in the fermentation broth (Figure 12). Thus, increase in concentration of delta-endotoxin followed the rate of increase in spore count.

2.2.3.2.5. Tx of fermented broth during the fermentation

Tx profiles during growth of Btk in three media are presented in Fig. 12. Tx value of fermented broth (containing delta-endotoxin, spores, viable cells, enzymes, cell debris, intracellular components, vegetative insecticidal proteins -VIPs, among others) was measured from 12h and it was found to increase until the end of fermentation (48h). Tx values at 48h of fermentation followed the order: SIW 30 g/L TS > SIW 15 g/L TS > synthetic medium (Table 12). The differences were statistically significant ($p \leq 0.05$).

Tableau 12. Maximum values of entomotoxicity of fermented broth, suspended pellet and liquid formulated products.

Tx ($x10^9$ SBU/L or BIU/L)	Synthetic medium	SIW 15 g/L TS	SIW 30 g/L TS
Max Tx of fermented broth ($x10^9$ SBU/L) [*]	11.7 ± 0.85^{c}	15.3 ± 1.0^{b}	17.8 ± 1.2^{a}
Max Tx of fermented broth (BIU/L) [**]	9.4-9.8 [c]	12.2-12.8 [b]	14.2-14.8 [a]
Max Tx of formulated product from fermented broth (BIU/L) [***]	7.7-8.0 [c]	10.0-10.5 [b]	11.7-12.2 [a]
Tx of suspended pellet ($x10^9$ SBU/L) [*]	22.7 ± 1.6^{ab}	20.9 ± 1.5^{b}	25.8 ± 1.8^{a}
Tx of suspended pellet (BIU/L) [**]	18.2-18.9 [ab]	16.7-17.4 [b]	20.6-21.5 [a]
Max Tx of formulated product from pellets re-suspension (BIU/L) [***]	14.9-15.5 [ab]	13.7-14.3 [b]	16.9-17.6 [a]

[*] The values are the mean of three determinations of two separate experiments conducted for each fermentation condition. The presented values are the mean ± SD. Different letters (stand near each value) within the same row indicate the significant differences among these values determined by one-factor analysis of variance (Turkey HSD test, $p \leq 0.05$).

[**] The Tx values (Billion Internation Units-BIU/L) calculated based on the Tx values ($x10^9$ SBU/L) with the BIU value less than SBU value 20-25% (Yezza et al., 2006).

[***] The Tx value (BIU/L) of formulated product calculated based on Brar et al., 2006b. The formulated product will reduce the Tx to approximately 22% of suspended pellet or fermented broth.

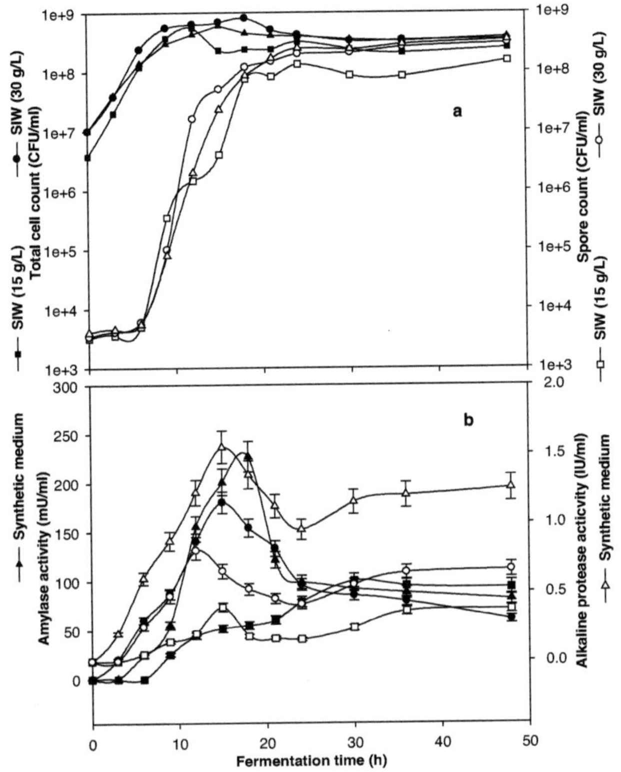

Figure 11. (a) Profiles of total cell count (CFU/ml) and spore count (CFU/ml); and (b) profiles of amylase activity and alkaline protease activity during fermentation using synthetic medium and SIW at 15 g/L TS and 30 g/L TS.

In general, entomotoxicity value is attributed to the concentration of main components, delta-endotoxin and spores [25]. Accordingly, Tx of synthetic medium fermented broth must have been higher than Tx of SIW 15 g/L TS fermented broth as delta-endotoxin and spore concentrations in synthetic medium fermented were considerably higher as compared to the values in

SIW 15 g/L fermented broth (Table 11). However, Tx was higher in 15 g/L TS SIW medium (Table 12). There could be two possible reasons for higher Tx value at lower spore count (SC) and delta-endotoxin concentration in SIW 15g/L TS fermented broth.

Firstly, Btk might have produced vegetative insecticidal proteins (Vips – a group of proteins produced during vegetative growth) which could act independently as biopesticide to kill target insect pests [26-28]. Vips production depends mainly on the type of components available in the fermentation media. Though, a synergistic action between Vips and delta-endotoxin against target insect pests is not yet fully explored, however, a specific Vip (Vip3A) having specific activity against lepidopteran insects produced by strain HD-1 of Btk (the same strain used in present research) in Peptone-Yeast extract medium [27] or in Terrific broth [28] has already been established. Therefore, Btk while growing in SIW would have produced Vip3A during the fermentation which would have acted independently synergistically with spores and delta-endotoxin (in fermented broth) to provide higher Tx value.

The second factor that might affect the Tx value is the production of Zwitttermicin (an antibiotic) during fermentation. Like Vips, Zwitttermicin synthesis also depends on the medium components [29]. It is known that Btk can produce Zwittermicin A [30]. This antibiotic which produced by *Bacillus cereus* UW85 can act synergistically with Btk in commercial biopesticide against Gypsy moth [31]. Production of non-identified antibiotic by Btk active against several human pathogens has been also detected in our laboratory (un-published results). Therefore, it is likely that Zwittermicin A could be produced in larger concentration during growth of Btk in SIW 15g/L than in synthetic medium and thus contributing towards the higher Tx value at low spore and delta-endotoxin concentration.

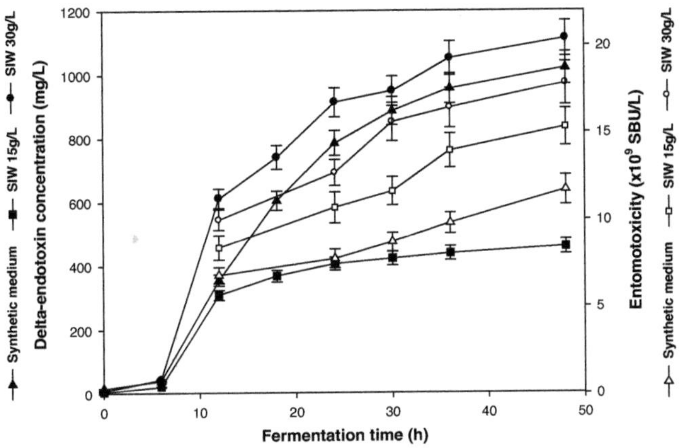

Figure 12. Delta-endotoxin concentration and entomotoxicity of synthetic medium and SIW fermented broth at 15g/L total solids and 30g/L total solids

2.2.3.3. Tx and field efficacy

Commercial Btk biopesticides products are normally based on International Unit (IU) determined using cabbage looper larvae as target insect for bioassay [21]. Field efficacy of Btk based biopesticides to specify application prescriptions for optimal protection of balsam fir, *Abies balsamea* (L.), healthy stands against the spruce budworm (*Choristoneura fumiferana)* was determined over a period of 6 years (1996-2001) in southwestern Québec [32]. At moderate larval densities (<30 larvae per 45-cm branch tip), similar foliage protection was achieved with one or two Btk applications of 30 BIU/ha (Billion International Units/ha). For larval densities more than 30 larvae per branch tip, two successive applications (with a gap of 5 or 10 days) of 30 BIU/ha significantly increased foliage protection. Increasing the application dosage from 30 to 50 BIU/ha did not lead to better foliage protection against high larval densities, but the current standard dosage of 30 BIU/ha saved more foliage than 15 BIU/ha against moderate populations. The recommended dosage of 30 BIU could be applied in lower application volumes (1.5 L/ha) by using a high-potency product (20 BIU/L) or higher application volumes (2.37L/ha) of a lower potency product (12.7 BIU/L) without affecting the field efficacy [32] Therefore, by estimating Tx of Btk fermented broth during the fermentation of different media, it was possible to decide if the formulated product could be produced with entomotoxicity ≥ 12.7 BIU/L directly from fermented broth without concentrating the spore and delta-endotoxin.

In our earlier research we found that SBU (spruce budworm unit) determined against spruce bud worm is 20-25% higher than IU [10]. Therefore, the SBU measured in fermentation broths for various media was converted to IU by using the above factor of conversion. The calculated values thus obtained are presented in Table 12. Further, to evaluate the potential of final formulated product from fermented broth for liquid formulation, the Tx values (BIU/L) were corrected (22% decrease) for dilution of Tx value during formulation (decrease in Tx due to dilution caused by addition of various ingredients to formulate the final product) [9] and results are presented in Table 12. Thus, among three cases, only fermented broth of SIW 30 g/L showed the potential for formulation (Tx 11.7-12.2 BIU/L) close to the potency (12.7 BIU/L) required for commercial products of low potency [32]. However, a word of caution regarding preparations of direct formulation from fermented broth require further verification in terms of economics due to requirement of large volume of storage and/or transportation.

2.2.3.4. Entomotoxicity of suspended pellets (spore and delta-endotoxin complex)

It has been specified that the main component for killing the target insect pest are delta-endotoxins which act specifically against insect pest and the spores may play a synergistic role in killing some target insect pests [25] Thus, for practical application, entomocidal potency or Tx of Bt based biopesticides is normally assessed based on the complex of spore and delta-endotoxin which is harvested by centrifugation of fermented broth [8,20]. Therefore, in the present case the fermented broths of 3 different media were centrifuged (9000 g, 30 min) and the obtained pellets were re-suspended with supernatant to obtain one-tenth volume of original fermented broth of each media and the suspension was subjected to bioassay to determine Tx value (SBU/L) against spruce budworm larvae which are summarized in Table 4. The difference of Tx value of synthetic medium with SIW (irrespective of TS concentration) is statistically insignificant ($p \leq 0.05$), however, there is significant difference between values of Tx obtained with SIW 30 and SIW 15 g/L TS ($p \leq 0.05$).

Further, the SBU values of re-suspended pellets were corrected for conversion to IU and dilution (22%) due fortification of ingredients (antimicrobial agents, preservatives, phagostimulants, stickers, etc.) for formulation [9]. The corrected values of IU for all media are presented in Table 4. The Tx values (BIU/L) were above the value (12.7 BIU/L) recommended for field application [32], irrespective of medium used for Btk production. Thus, the formulated product can be effectively used for the control of spruce budworm. Based on standard application of 30 BIU/ha, the calculated volume of formulated product required will be 2.1-2.2L, 1.7-1.8L and 1.9-2.0 L per hectare and would be produced from 21-22L, 17-18L and 19-20L of fermented broth for SIW 15 g/L, SIW 30g/L and synthetic medium, respectively.

In fact, the raw material accounts for ~ 40 % of total biopesticide production cost [33]. Therefore, using alternative raw materials with low cost (or negative cost) is very important. Thus, SIW is a source of nutrients that require less cost input, demonstrated a very high potential for commercial exploitation/competition and is very important for those countries where the sources of SIW are abundant and facing with acute wastewater treatment problems.

2.2.4. Conclusion

Following conclusions could be drawn from the foregoing research are:

➢ Optimal total solids concentration of SIW and inoculum volume of pre-culture for Btk based biopesticides production were found to be 30 g/L and 2.5% (v/v), respectively.

➢ Higher cell count, spore count, amylase activity, alkaline protease activity, delta-endotoxin concentration and entomotoxicity were obtained in the case of optimal TS of SIW 30 g/L as compared to SIW 15 g/L TS.

➢ Maximum entomotoxicity value of Btk fermented broth was observed in SIW 30 g/L and followed the order, SIW 30 g/L TS > SIW 15 g/L TS > synthetic medium.

➢ A low spore count and delta-endotoxin concentration was observed in SIW 15g/L TS than semi-synthetic medium; however, SIW 15 g/L TS revealed higher entomotoxicity value.

➢ Entomotoxicity value of suspended pellet of SIW 30 g/L TS revealed higher potential for application and forestry protection against spruce budworm larvae as compared to entomotoxicity values in suspended pellets of synthetic medium and SIW 15 g/L TS.

2.2.5. Acknowledgements

The authors are sincerely thankful to the Natural Sciences and Engineering Research Council of Canada (Grants A4984, STP235071, Canada Research Chair) for financial support. The views or opinions expressed in this article are those of the authors and should not be construed as opinions of the U.S. Environmental Protection Agency.

2.2.6. References

1. Goldberg I., Sneh B., Battat E., Klein D., Optimisation of a medium for a high yield production of spore-crystal preparation of *Bacillus thuringiensis* effective against the Egyptian cotton leaf warm *Spodoptera littoralis* Boisd, *Biotechnol. Lett.*, **2**, 419-426 (1980).

2. Scherrer P., Luthy P., Trumpi B., Production of δ-Endotoxin by *Bacillus thuringiensis* as a function of glucose concentrations, *Appl. Microbiol.*, **25**, 644-646 (1973).

3. Arcas J., Yantorno O.M., Ertola R.J., Effect of high concentration of nutrients on *Bacillus thuringiensis* cultures, *Biotechno. Lett.*, **9**, 105–110 (1987).

4. Farrera R.R., Pérez-Guevara F., de la Torre M., Carbon: nitrogen ratio interacts with initial concentration of total solids on insecticidal crystal protein and spore production in *Bacillus thuringiensis* HD-73, *Appl. Microbiol. Biotechnol.*, **49,** 758-765 (1998).

5. Lachhab K., Tyagi R.D., Valéro J.R., Production of *Bacillus thuringiensis* biopesticides using wastewater sludge as a raw material: effect of inoculum and sludge solids concentration, *Process Biochem.*, **37,** 197-208 (2001).

6. Vidyarthi A.S., Tyagi R.D., Valéro J.R., Surampalli R.Y., Studies on the production of *B. thuringiensis* based biopesticides using wastewater sludge as a raw material, *Water Res.*, **36**, 4850–4860 (2002).

7. Brar S.K., Verma M., Tyagi R.D., Valéro J.R., Surampalli R.Y., Starch industry wastewater-based stable *Bacillus thuringiensis* liquid formulations, *J. Econ. Entomol.*, **98**, 1890-1898 (2005).

8. Brar S.K., Verma M., Tyagi R.D., Valéro J.R., Surampalli R.Y. Efficient centrifugal recovery of *Bacillus thuringiensis* biopesticides from fermented wastewater and wastewater sludge, *Water Res.*, **40**, 1310-1320 (2006a).

9. Brar S.K., Verma M., Tyagi R.D., Valero J.R., Surampalli R.Y., Techno-economic analysis of *Bacillus thuringiensis* production process, Final report to INRS-ETE, Research report No. R-892 (2006b).

10. Yezza A., Tyagi R.D., Valéro J.R., Surampalli R.Y., Bioconversion of industrial wastewater and wastewater sludge into *Bacillus thuringiensis* based biopesticides in pilot fermentor, *Biores. Technol.*, 97, 1850-1857 (2006).

11. APHA, AWWA, WPCF, Standard methods for examination of water and wastewaters, 20th ed, American Public Health Association, Washington, DC, (1998).

12. Avignone-Rossa C., Arcas J., Mignone C., *Bacillus thuringiensis*, sporulation and δ-endotoxin production in oxygen limited and nonlimited cultures, *World J. Microbiol. Biotechnol.*, **8**, 301–304 (1992).

13. Aiba S., Humphrey A.E., Millis N.F., Biochemical Engineering, 2nd ed, Academic Press, New York, (1973).

14. Stanbury P.F., Whitaker A., Hall S.J., Principles of fermentation technology, 2nd edition, Elsevier Science Ltd., U.K., (2003).

15. Kunitz M., Crystalline soybean trypsin inhibitor, *J. General Physiol.*, **30**, 291–310 (1947)

16. Miller G., Use of dinitrosalicylic acid reagent for determination of reducing sugar, *Analytical Chem.* **31**, 426-428 (1959).

17. Zouari N., Jaoua S., Production and characterization of metalloproteases synthesized concomitantly with delta-endotoxin by *Bacillus thuringiensis* subsp. *kurstaki* strain grown on gruel-based media, *Enz. Microbial. Technol.*, **25**, 364–371 (1999a).

18. Zouari N., Jaoua S., The effect of complex carbon and nitrogen, salt, Tween-80 and acetate on delta-endotoxin production by a *Bacillus thuringiensis* subsp. *Kurstaki*, *J. Ind. Microbiol. Biotechnol.*, **23**, 497-502 (1999b).

19. Bradford M.M., A rapid and sensitive for the quantization of microgram quantities of protein utilizing the principle of protein-dye binding, *Analytical Biochem.*, **72**, 248-254 (1976).
20. Dulmage H.T., Correa J.A., Martinez A.J., Coprecipitation with lactose as a means of recovering the spore-crystal complex of *Bacillus thuringiensis*, *J. Invertebr. Pathol.*, **15**, 15-20 (1970).
21. Dulmage H.T., Boening O.P., Rehnborg C.S., Hansen G.D., A proposed standardized bioassay for formulations of *Bacillus thuringiensis* based on the international unit, *J. Invertebr. Pathol.*, **18**, 240-245 (1971).
22. Beegle C.C., Bioassay methods for quantification of *Bacillus thuringiensis* δ-endotoxin, analytical chemistry of *Bacillus thuringiensis*, in *Analytical chemistry of Bacillus thuringiensis*, Hickle L.A., Fitch W.L., eds, USA: American Chemical Society, ISBN 0841218153, pp. 14–21 (1990).
23. Baum J.A., Malvar T., Regulation of insecticidal crystal protein production in *Bacillus thuringiensis*, *Mol. Microbiol.*, **18**, 1-12 (1995).
24. Bechtel D.B., Bulla L.A. Jr., Electron microscope study of sporulation and parasporal crystal formation in *Bacillus thuringiensis*, *J. Bacteriol.*, **127**, 1472-1481 (1976).
25. Schnepf E., Crickmore N., van Rie J., Lereclus D., Baum J., Feitelson J., Zeigler D.R., Dean D.H., *Bacillus thuringiensis* and its pesticidal crystal proteins, *Microbiol. Mol. Biol. Rev.*, **62**, 775-806 (1998).
26. Estruch J.J., Warren G.W., Mullins M.A., Nye G.J., Craig J.A., Koziel M.G., Vip3A, a novel *Bacillus thuringiensis* vegetative insecticidal protein with a wide spectrum of activities against lepidopteran insects, *Proc. Nat. Acad. Sci. USA*, **93**, 5389-5394 (1996).
27. Donovan W.P., Donovan J.C., Engleman J.T., Gene knockout demonstrates hat *vip3A* contributes to the pathogenesis of *Bacillus thuringiensis* toward *Agrotis ipsilon* and *Spodoptera frugiperda*, *J. Invertebr. Pathol.*, **78**, 45–51 (2001).

28. Milne R., Liu Y., Gauthier D., van Frankenhuyzen K., Purification of Vip3Aa from *Bacillus thuringiensis* HD-1 and its contribution to toxicity of HD-1 to spruce budworm (*Choristoneura fumiferana*) and gypsy moth (*Lymantria dispar*) (Lepidoptera), *J. Invertebr. Pathol.*, Doi: 10.1016/j.jip.2008.05.002 (2008).

29. Milner J.L., Raffel S.J., Lethbridge B.J., Handelsman J., Culture conditions that influence accumulation of zwittermicin A by *Bacillus cereus* UW85, *Appl. Microbiol. Biotechnol.*, **43**, 685-691 (1995).

30. Stabb E.V., Jaconson L.M., Handelsman J., Zwittermicin A-producing strains of Bacillus cereus from diverse soils, *Appl. Environ. Microbiol.*, **60**, 4404-4412 (1994).

31. Broderick N.A., Goodman R.M., Raffa K.F., Handelsman J., Synergy between Zwittermicin A and *Bacillus thuringiensis* subsp. *kurstaki* against Gypsy Moth (Lepidoptera: Lymantriidae), *Environ. Entomol.*, **29**, 101-107 (2000).

32. Bauce E., Carisey N., Dupont A., van Frankenhuyzen K., *Bacillus thuringiensis* subsp. *kurstaki* aerial spray prescriptions for balsam fir stand protection against spruce budworm (Lepidoptera: Tortricidae), *J. Econ. Entomol.*, **97**, 1624-1634 (2004).

33. Lisanky S.G., Quinlan R., Tassoni G., The *Bacillus thuringiensis* production handbook, CPL Press, Newbury, UK (1993).

Partie 2.3. Production de biopesticide à base de *Bacillus thuringiensis* (Btk HD-1) en utilisant des eaux usées d'industrie d'amidon (SIW) enrichies de différentes sources de carbone et d'azote comme milieux de fermentation

Résumé

Différentes sources de carbone et d'azote ont été ajoutées dans des SIW à diverses concentrations et ces mélanges ont été utilisés comme milieux pour la production en erlenmeyers de biopesticides à base de Btk HD-1. Les concentrations en cellules, en spores et en delta-endotoxines (dans des bouillons fermentés de 48h) ont été évaluées pour déterminer quels types de sources de carbone/azote et à quelles concentrations doivent être ajoutés dans les SIW pour obtenir des concentrations élevées des paramètres cités précédemment. L'amidon de maïs (comme source de carbone), ajoutée aux SIW à une concentration finale de 1,25%, p/v a conduit à obtenir des concentrations plus élevées en spores et en delta-endotoxines par rapport aux SIW contenant d'autres concentrations d'amidon de maïs ou d'autres sources de carbone. Les SIW fortifiées avec du Tween 80 à 0,3 ou 0,4%, v/v ont donné la plus grande concentration en delta-endotoxines par rapport aux SIW complétées avec d'autres concentrations de ce Tween. Dans le cas d'apport d'azote, les SIW complétées avec différentes sources azotées n'ont pas permis d'augmenter significativement les concentrations en cellules, en spores et en delta-endotoxines, l'ajout de cet élément a même entraîné des résultats négatifs en concentrations plus élevées aux SIW. En outre, les SIW complétées avec deux sources de carbone soit l'amidon de maïs à 1,25%, p/v et le Tween 80 à 0,2 ou 0,3%, v/v ont permis d'atteindre des concentrations plus élevées en cellules, en spores et en delta-endotoxines par rapport à ces paramètres obtenus dans les SIW complétées avec l'amidon de maïs à 1,25% (w / v) et le Tween 80 à 0,1 ou 0,4%, v/v. La production de Btk HD-1 par fermentation en utilisant des SIW complétées avec l'amidon de maïs à 1,25%, p/v et le Tween 80 à 0,2%, v/v (désignée comme SIWST) a été effectuée en bioréacteur de 15 L dans des conditions contrôlées de pH, d'oxygène dissous et autres paramètres. L'entomotoxicité du bouillon fermenté et celle de la suspension du culot des SIWST étaient plus élevées de 16,0% et 24,9% de plus que celles du bouillon fermenté et de la suspension en culot des SIW (le contrôle), respectivement.

Bacillus thuringiensis based biopesticides production using starch industry wastewater fortified with different carbon/nitrogen sources as fermentation media

Abstract

Starch industry wastewater (SIW) supplemented with different concentration of carbon (glucose and corn starch), nitrogen (beef extract, bactopeptone, peptone from casein, yeast extract) and Tween 80 was used as media for the production of *Bacillus thuringiensis* var. *kurstaki* HD-1 (Btk) biopesticides in shake flask fermentation. The total cell count, spore count and delta-endotoxin concentration of fermented broths at 48h were assayed to determine optimum concentration of carbon and nitrogen sources and Tween 80. Cornstarch (as carbon source) supplemented at final concentration of 1.25 % w/v gave highest cell count, spore count and delta-endotoxin concentration. SIW supplemented with different nitrogen sources did not cause a significant increase in cell, spore and delta-endotoxin synthesis but it gave negative results when fortified into SIW at higher concentration. Further, SIW supplemented with both cornstarch at 1.25 %, w/v and Tween 80 at 0.2 or 0.3 %, v/v gave the higher cell, spore and delta-endotoxin concentrations as compared to these parameters obtained from SIW supplemented with both cornstarch at 1.25% (w/v) and Tween 80 at 0.1 or 0.4 %, v/v. Fermentation of Btk using SIW supplemented with both cornstarch at 1.25 %, w/v and Tween 80 at 0.2 %, v/v (designed as SIWST) in 15 L fermentor was conducted. Entomotoxicity of fermented broth and suspended pellet of SIWST was 16.0 % and 24.9 % higher than entomotoxicity of fermented broth and suspended pellet of SIW, respectively.

Key words: *Bacillus thuringiensis* var. *kurstaki* HD-1, delta-endotoxin, entomotoxicity, carbon and nitrogen sources, starch industry wastewater

2.3.1. Introduction

Production of *Bacillus thuringiensis* (Bt) based biopesticide requires the optimisation of fermentation medium which should consists of suitable carbon/nitrogen sources as well as other minerals (Zouari et al. 1998). Many studies on optimization of medium composition employing semi-synthetic medium or alternative cheap medium (crude gruel and fishmeal) have been reported (Scherrer et al. 1973; Liu and Bajpai 1995; Zouari et al. 1998; Zouari and Jaoua 1999). In case of wastewater sludge use as a raw material for Bt production, it was demonstrated that supplementation of glucose or yeast extract could enhance significant cell, spore production and entomotoxicity (Tx) of the fermented broth (Leblanc 2003). Recently, starch industry wastewater (SIW) has been used as a raw material for Bt based biopesticide production and the results demonstrated net advantage as compared to semi-synthetic medium in many respects: production of value added product and concomitantly wastewater treatment, very low cost of raw material and higher biopesticidal potential compared to that obtained on semi-synthetic medium (Yezza et al. 2006; Vu et al. 2008; 2009). Our previous research demonstrated that use of optimum total solids (TS) concentration of SIW (30 g/L) for Bt based biopesticide production significantly enhanced the cell and spore counts, delta-endotoxin concentration and Tx of the fermented broth as well as Tx of suspended pellet (10 times concentrated fermented broth) (Vu et al. 2009). The forgoing research therefore indicated that SIW at low TS should be supplied with more nutrients (under carbon or nitrogen sources) to satisfy the requirement of Btk metabolism for production of higher spore and delta-endotoxin concentration and higher Tx value. Therefore, the purpose of this research was to investigate the effect of fortification of different carbon and/or nitrogen sources to SIW on Btk cell, spore, delta-endotoxin concentrations and biopesticidal potential against spruce budworm.

2.3.2. Materials and Methods

2.3.2.1. Bacterial strain and biopesticide production medium

Bacillus thuringiensis var. *kurstaki* HD-1 (ATCC 33679) (Btk) was used in this study. The Btk was subcultured and streaked on tryptic soya agar (TSA) plates, incubated for 48 hour at 30 ± 1°C and then preserved at 4.0 ± 1 °C for future use (Vu et al. 2008).

During the course of this study, starch industry wastewater (SIW) from ADM-Ogilvie (Candiac, Québec, Canada) was used as a raw material for Btk growth. Total solids (TS), volatile solids (VS), suspended solids (SS) and volatile suspended solids (VSS), pH, total carbon (C_t), total Kjeldahl nitrogen (N_t), ammonia nitrogen ($N–NH_4^+$), total phosphorus and metals concentration (Al, Ca, Cd, Cr, Cu, Fe, Mg, Mn, Na, Ni, Pb and Zn) were determined according to Standard Methods (APHA et al. 1998). Physico-chemical characteristics of SIW are presented in Table 13.

Synthetic medium (semi-synthetic medium) based on soybean meal was also used as control medium. It consist of (g/L): soybean meal (15.0); glucose (5.0); starch (5.0), K_2HPO_4 (1.0); KH_2PO_4 (1.0); $MgSO_4.7H_2O$ (0.3); $FeSO_4.7H_2O$ (0.02); $ZnSO_4.7H_2O$ (0.02); $CaCO_3$ (1.0).

2.3.2.2. Inoculum preparation

All the media used for inoculum preparation were adjusted to pH 7 before autoclaving. A loopful of Btk from TSA plate was used to inoculate a 500-ml Erlenmeyer flask containing 100 ml of sterilised tryptic soya broth (TSB) medium. The flask was incubated on a rotary shaker at 250 revolutions per min (rpm) and 30 °C for 8–12 h. A 2% (v/v) inoculum from this flask was then used to inoculate 500-ml Erlenmeyer flasks containing 100 ml of sterilised starch industry wastewater (SIW) and incubating for 8-12 h. Then, the actively growing cells from these flasks were used as an inoculum (pre-culture) for the production of Btk-based biopesticide in shake flasks each containing 100 ml SIW supplemented with different carbon and nitrogen sources at different final concentrations.

Tableau 13. Characteristics of starch industry wastewater (SIW)

Parameter	Concentration* (g/L)	Parameter	Concentration* (mg/kg TS)
Total solids [TS]	16.5 ± 0.8	Al	203.8 ± 6.5
Volatile solids [VS]	11.5 ± 0.7	Cd	0.14 ± 0.02
Suspended solids [SS]	6.2 ± 0.3	Cr	5.80 ± 0.04
Volatile suspended solids [VSS]	6.0 ± 0.3	Cu	22.70 ± 0.27
pH	3.5 ± 0.1	Fe	350.7 ± 14.8
	Concentration* (g/kg TS)	Mn	47.50 ± 0.30
C_t	434 ± 8	Ni	2.40 ± 0.02
N_t	34 ± 1	Pb	1.20 ± 0.02
$N\text{-}NH_3$	0.70 ± 0.01	Zn	70.20 ± 0.65
P_t	7.0 ± 0.2		
$P\text{-}PO_4^{3-}$	4.00 ± 0.02		
Ca	4.00 ± 0.07		
K	13.0 ± 0.2		
Mg	3.00 ± 0.03		
Na	90.0 ± 0.3		
S	6.0 ± 0.1		

* The presented values are the mean ± standard deviation (SD).

2.3.2.3. Shake flask fermentation

Erlenmayer flasks each containing 100 ml of SIW (Total solid concentration of 16.5 g/L) were supplemented with different carbon sources: glucose, cornstarch at concentrations of 0.10, 0.25, 0.50, 1.0 %, w/v or Tween 80 at 0.0, 0.1, 0.2, 0.3, 0.4 % v/v. Similarly, Erlenmeyer flasks each containing 100 ml of SIW were supplemented with different nitrogen sources (beef extract, bactopeptone, peptone from casein, yeast extract) at concentrations of 0.00; 0.10; 0.25; 0.50; 1.0 %, w/v. All these flasks were adjusted to pH 7.0, sterilized at 121 °C for 20 min and were left at room temperature for cooling, then inoculated with 2% (v/v) of pre-culture prepared as above. These flasks were incubated in a shaker-incubator (New Brunswick) for 48 h at 30°C and 220 rpm. Samples were collected at the end of fermentation process (48h) to determine the total cell count (TC), viable spore count (VS) and delta-endotoxin concentration as described in following sections. The carbon/nitrogen source which enhanced significantly spore and delta-endotoxin concentration as compared to SIW control (without any fortification) was further investigated in bioreactor.

2.3.2.4. Fermentation procedure in 15-L computer controlled bioreactor

To evaluate the effects carbon/nitrogen fortification on Btk based biopesticides production, experiments in bioreactor were conducted using three media: synthetic medium, SIW without fortification any carbon/nitrogen sources and SIW fortified with optimized concentration of carbon/nitrogen source chosen from shake flask experiments. Fermentation was carried out in a stirred tank 15 L fermentor (working volume: 10 L, Biogénie, Que., Canada) equipped with accessories and programmable logic control (PLC) system for dissolved oxygen (DO), pH, anti-foam, impeller speed, aeration rate and temperature. The software (iFix 3.5, Intellution, USA) allowed automatic set-point control and integration of all parameters via PLC. The fermentor was then inoculated with 2% v/v inoculum aseptically with preculture of Btk in exponential phase (8-12 h). In order to keep the DO above 25% saturation, air flow rate and agitation rate were varied between 2-4 LPM and 250–500 rpm, respectively. This ensured the critical DO level for Bt above 25 % (Avignone-Rossa et al. 1992). The temperature was maintained at 30 ± 1 °C by circulating water through the fermentor jacket. Fermentation pH was controlled automatically at 7 ± 0.1 through computer-controlled peristaltic pumps by addition of pH control agents: NaOH 4 M/H_2SO_4 3 M (sodium hydroxide 4 M/sulphuric acid 3 M). Dissolved oxygen and pH were continuously monitored by means of a polarographic dissolved oxygen probe and of a pH sensor (Mettler-Toledo, USA), respectively. Samples were collected periodically to monitor the changes in TC, SC, delta-endotoxin concentration and Tx as described in following sections.

2.3.2.5. Analysis of parameters

2.3.2.5.1. Determination of volumetric oxygen transfer coefficient (K_La), Oxygen Transfer Rate (OTR) and Oxygen Uptake Rate (OUR)

The volumetric oxygen transfer coefficient (K_La) measurement was based on the dynamic method (Aiba et al. 1973; Stanbury et al. 2003). This technique consisted in interrupting the air input. Afterwards, the aeration was re-established. The decrease (aeration off) and the increase (aeration on) in DO concentration were recorded. K_La was determined from the mass balance on DO just after each sampling of the fermentation broth. During batch fermentation, the mass balance for the DO concentrations could be written as follows: $dC_L/dt = OTR - OUR$

OTR: represents the oxygen transfer rate from the gas phase to the liquid phase:

$OTR = K_La(C^*-C_L)$

OUR: represents the oxygen uptake rate:

$OUR = Q_{O2}X$

K_La: volumetric oxygen transfer coefficient

C^*: oxygen concentration at saturation

C_L: Dissolved oxygen concentration in the medium

Q_{O2} : specific oxygen uptake rate

X: cell concentration

Oxygen concentration in the fermentation broth was converted from % air saturation to mmol O_2/l as follows: the DO electrode was calibrated in the medium at 30 ± 1 °C and then transferred to air-saturated distilled water at known temperature and ambient pressure. This reading was used with the known saturation concentration of oxygen at saturation in distilled water (0.07559 mmol/l) (100%), to estimate the saturation concentration of oxygen at saturation in the cultivation media at 30 °C.

2.3.2.5.2. Estimation of delta-endotoxin concentration produced during the fermentation

Delta-endotoxin concentration was determined based on the solubilization of insecticidal crystal proteins in alkaline condition (Zouari and Jaoua 1999): 1 ml of samples collected during fermentation was centrifuged at 10 000 g for 10 min. at 4°C. The pellet containing a mixture of spores, insecticidal crystal proteins, cell debris and residual suspended solids was used to estimate the concentration of alkaline solubilised insecticidal crystal proteins (delta-endotoxin): These pellets were washed three times with 1 ml of 0.14 M NaCI - 0.0 1 % Triton X- 100 solution. The washing helped in eliminating the soluble proteins and proteases which might have adhered on the pellets and could affect the integrity of the crystal protein. The insecticidal crystal proteins in the pellet were solubilized with 0.05N NaOH (pH 12.5) for three hours at 30°C with stirring. The suspension was centrifuged at 10 000 g for 10 min. at 4°C and the pellet, containing spores, cell debris and residual suspended solids was discarded. The supernatant, containing the alkaline solubilised insecticidal crystal proteins was used for determination of delta-endotoxin concentration by Bradford method (Bradford, 1976) using bovine serum albumin as standard protein. The values presented were the average results (± SD) of triplicates of two independent experiments.

2.3.2.5.3. Estimation of total count (TC) and spore count (SC)

The TC and SC were performed by counting colonies grown on TSA medium (Yezza et al. 2006). For all counts, the average of at least triplicate plates was used for each tested dilution. For enumeration, 30–300 colonies were enumerated per plate. The results were expressed as colony forming units per mL (CFU/ml).

2.3.2.5.4. Preparation of pellet for bioassay

At the end of fermentation, fermented broth of three media were adjusted to pH 4.5 and centrifuged at 9000 g for 30 min (Brar et al. 2006). The individual pellets (spore and delta-endotoxin complex) so obtained were re-suspended with supernatant resulting in one-tenth volume of original fermented broth. The suspended pellets were used to determine Tx through bioassay.

2.3.2.5.5. Bioassay technique

The potential Tx of Btk HD-1 was estimated by bioassay against third instar larvae of spruce budworm (Lepidoptera : *Choristoreuna fumiferena*) following the diet incorporation method (Dulmage et al. 1971; Beegle 1990). The commercial preparation 76B FORAY (Abbott Laboratory, Chicago, IL) was used as a reference standard to analyse the Tx. The detail procedure of bioassay technique was presented in Yezza et al. (2006). The Tx was evaluated by comparing the mortality percentages of dilutions of the samples with same dilutions of the standard. In our research, Tx was expressed as relative spruce budworm units (SBU/L). On comparison of Tx of Bt-fermented samples, it was found that SBU reported in our study was 20–25% higher than IU (Yezza et al. 2006). The presented values are the mean of three determination of two independent experiments ± SD.

2.3.3. Results and Discussion

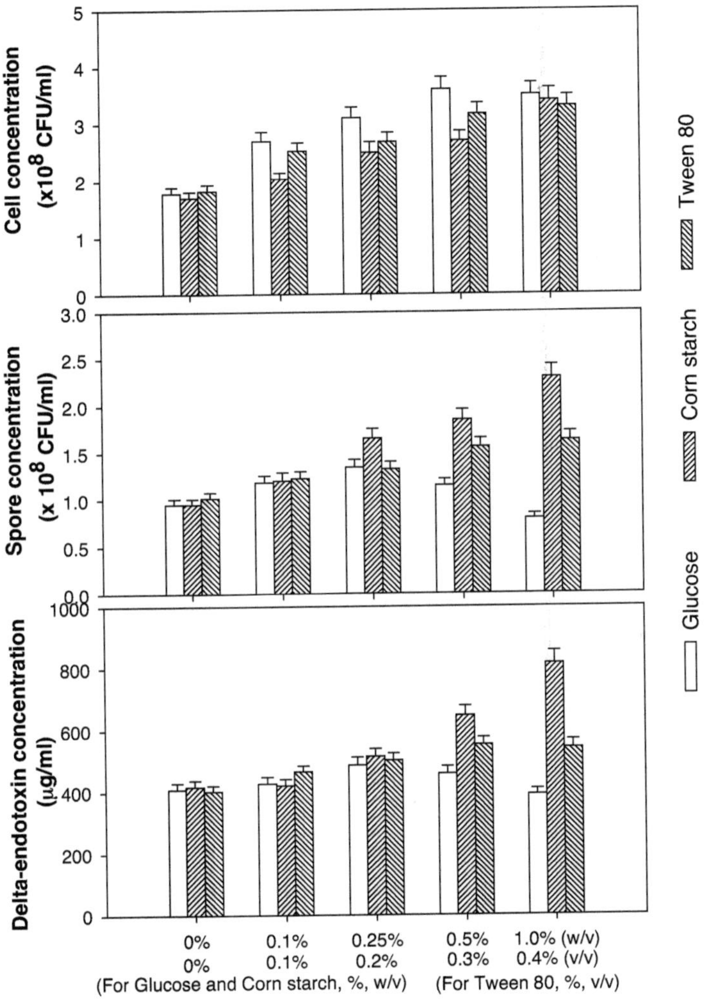

Figure 13. Effects of different concentration of carbon sources fortified into SIW on the (a) cell concentration, (b) spore concentration and (c) delta-endotoxin concentration

2.3.3.1. Shake flask experiment

2.3.3.1.1. Effects of SIW fortification with different carbon sources on growth and delta-endotoxin synthesis of Btk HD-1

Effects of carbon sources (glucose and corn starch) added to SIW at different final concentrations (w/v) on total cell count (TC), spore count (VS), delta-endotoxin concentration are illustrated in Figure 13. Maximum TC concentration was obtained at glucose concentration of 0.50 % w/v, however, maximum VS and delta-endotoxin concentrations were obtained at glucose concentration of 0.25 % w/v (Figure 13). Higher glucose concentration (0.5-1.0 %, w/v) in the medium resulted in a decrease of VS and delta-endotoxin concentration. A decrease in these parameters at high glucose concentrations could be due to catabolic repression (Zouari et al. 1998). Catabolic repression is a decrease in the activity of certain auxiliary catabolic enzymes when an easily metabolizable substrate is available in the growth medium. Commonly, this effect is caused by glucose (glucose repression) or by metabolites produced from glucose. Further, glucose added to SIW at higher concentration may cause the production of excess quantity of organic acids which could be inhibitory to growth (Scherrer et al. 1973). In case of cornstarch, 1.0 % w/v concentration gave the highest TC, VS and delta-endotoxin concentration (Figure 13) and their values were higher compared to glucose. In fact, cornstarch is a popular carbon source in the semi-synthetic media to produce biopesticides by Bt or proteases by *Bacillus licheniformis* (Yezza et al. 2006). Cornstarch is a suitable carbon source for Bt fermentation due to the following reasons: (1) it is a complex substrate (carbohydrate) which can overcome the catabolic repression (Zouari and Jaoua 1999); (2) Bt can synthesize amylases to hydrolyse starch into glucose for growth (Zouari and Jaoua 1999; Vu et al. 2008; 2009). Our results were also in agreement with other reports (Zouari and Jaoua 1999) that the use of complex substrate overcome the catabolic repression and furnish higher concentration of spores and delta-endotoxin.

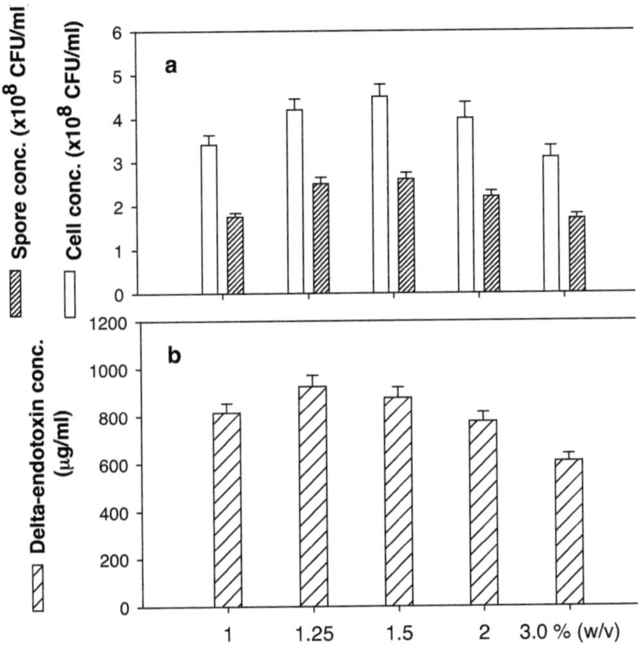

Figure 14. Effects of different concentrations (1.0; 1.25; 1.5; 2.0; and 3.0 %, w/v) of corn starch fortified into SIW on (a) cell and spore concentration and (b) delta-endotoxin concentration

However, this experiment was only conducted up to maximum concentration of cornstarch 1.0 % w/v and it was not known if increase in concentration of cornstarch in SIW could further enhance cell, spore and delta-endotoxin concentrations produced by Btk. Therefore, other experiment was conducted in which cornstarch was added to SIW to obtain higher final concentrations such as 1.25; 1.5; 2.0 and 3.0 %, w/w and these mixtures were used as fermentation media for Btk. Results revealed that cornstarch added to SIW at concentration of 1.5 % (w/v) gave better results in term of TC, however, VS and delta-endotoxin were almost same at cornstarch concentrations of

1.25 % and 1.50 %, w/v (Figure 14). Therefore, cornstarch concentration of 1.25 % w/v was selected for further study.

Increase of Tween 80 concentration in SIW (from 0.1 to 0.4%, v/v) caused an increase in TC, VS and delta-endotoxin concentration in comparison to the control (without Tween 80) (Figure 13). However, TC, VS and delta-endotoxin concentrations obtained at Tween 80 concentrations of 0.3 or 0.4 %, v/v were not significantly different (Figure 13). It is known that Tween 80 is a nonionic surfactant and emulsifier derived from polyethoxylated sorbitan and oleic acid and is often used in foods and pharmaceutical industry. Tween 80 plays an important role of facilitating the passage of soluble proteins in bacterial cells and increases the assimilation of the nutrients by bacteria (Singh et al. 2007). Similar results were also obtained in the fermentation of Btk using wastewater sludge fortified with Tween 80 at a concentration of 0.2 % (v/v) (Brar et al. 2005). Fortification of Tween 80 (0.1% v/v) into crude gruel and fishmeal medium resulted in a considerable increase in growth and delta-endotoxin synthesis (Zaouri and Jaoua, 1999). It is recognized that Tween 80 can act as a carbon source as well as a surface active compound, which enhance the transfer of oxygen in fermentation (Zaouri and Jaoua 1999; Brar et al. 2005). Moreover, Tween 80 can also play an important role for formulation of final Bt product (Brar et al. 2005).

It is interesting to note that the optimum Tween 80 concentration is different for different media. Crude gruel and fishmeal medium showed optimum Tween 80 concentration of 0.1% v/v (Zaouri and Jaoua 1999), whereas in wastewater sludge 0.2% v/v concentration resulted in the highest Tx value (Brar et al. 2005). In this research, 0.3% or 0.4 % v/v Tween 80 was found to be the suitable to augment TC, VS and delta-endotoxin synthesis by Btk. The higher requirement of Tween 80 concentration for SIW as compared to other raw materials (sludge or crude gruel and fishmeal) could be due to the fact that SIW might not contain enough carbon sources and it was probably used by Btk as a carbon source.

2.3.3.1.2. Effets of fortification of different nitrogen sources to SIW on growth and delta-endotoxin synthesis of Btk HD-1

Effects of different nitrogen sources (bactopeptone, beef extract, peptone from casein and yeast extract) fortified into SIW at different concentrations on TC, VS and delta-endotoxin concentration are illustrated in Figure 15. Bactopeptone, peptone from casein and yeast extract added to SIW at 0.25 %, w/v caused an increase in TC, however, the increase was not significant as compared to the control (Figure 15). Moreover, fortification of these nitrogen sources (bactopeptone, peptone from casein and yeast extract) to SIW decreased VS and delta-endotoxin concentration. Beef extract fortified into SIW at concentration of 0.50% w/v gave significantly higher TC as compared to the control, however, VS and delta-endotoxin concentrations were significantly higher at 0.1 %w/v concentration of beef extract. The negative effect of fortification of these nitrogen sources (bactopeptone, peptone from casein and yeast extract) to SIW on VS and delta-endotoxin synthesis by Btk could be due to the following reasons: (1) SIW may already possess enough nitrogen source in inorganic form ($NH4^+$) (Table 13). In fact, our previous research also illustrated that use of NH_4OH as pH control agent in combination with either H_2SO_4 or CH_3COOH during fermentation of Btk using SIW as raw material had negative effect on TC, VS and delta-endotoxin production (Vu et al. 2008); (2) It is known that the first step in amino acid assimilation is deamination (the removal of amino group from an amino acid) and it is also known that excess nitrogen from deamination may be excreted as ammonium ion, thus making pH medium alkaline (Prescott et al. 2002). The alkaline medium might cause inhibition to cell growth and finally the sporulation and delta-endotoxin synthesis by Btk. Therefore, fortification of nitrogen sources into SIW for Btk based biopesticide production is not required.

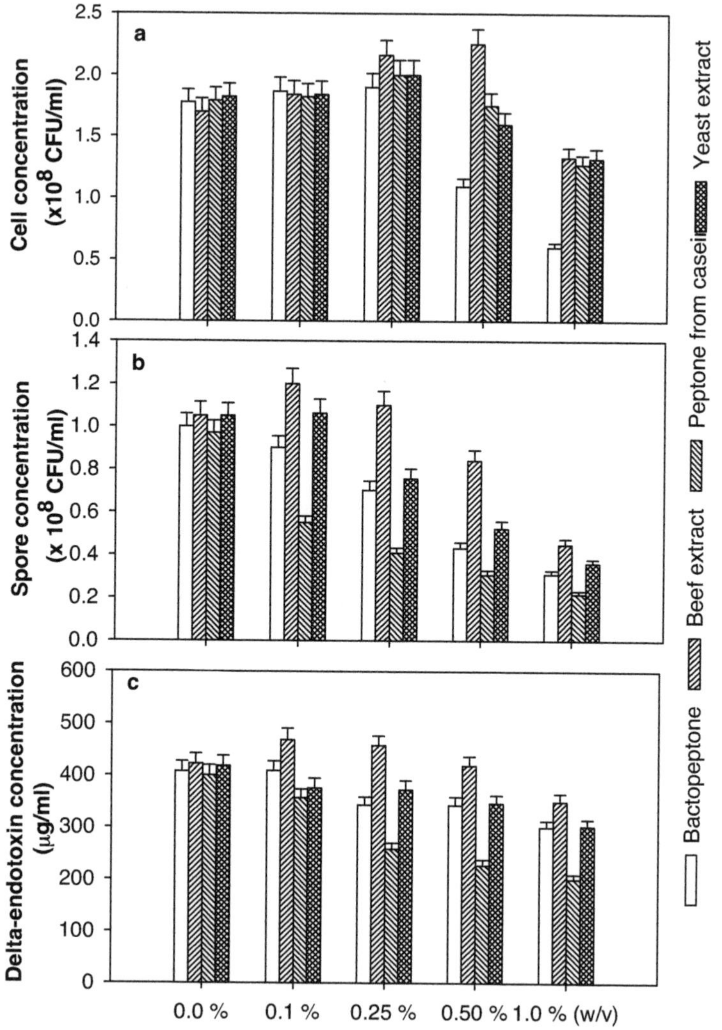

Figure 15. Effects of different concentration of nitrogen sources fortified into SIW on (a) cell concentration; (b) spore concentration and (c) delta-endotoxin concentration

2.3.3.1.3. Fortification of both cornstarch and Tween 80 into SIW

Tween 80 concentration was varied from 0.1 to 0.4 %, v/v while cornstarch concentration kept constant at 1.25 %w/v. It was found that Tween 80 concentration of 0.2 or 0.3 % v/v gave almost same concentrations of TC, VS and delta-endotoxin but higher than Tween 80 concentration of 0.1 or 0.4 % (Figure 16). Also, cornstarch concentration of 1.25 % w/v and Tween 80 of 0.2 %v/v gave delta-endotoxin concentration of 1033.4 µg/ml which was 2.5 times higher than that in control without fortification (415.5 µg/ml). Therefore, fortification of Tween 80 at concentration of 0.2 %v/v in combination with corn starch concentration of 1.25% w/v or SIWS (SIW + cornstarch at 1.25 %, w/v) was chosen for further experiments in bioreactor at controlled pH and dissolved oxygen (DO) concentration.

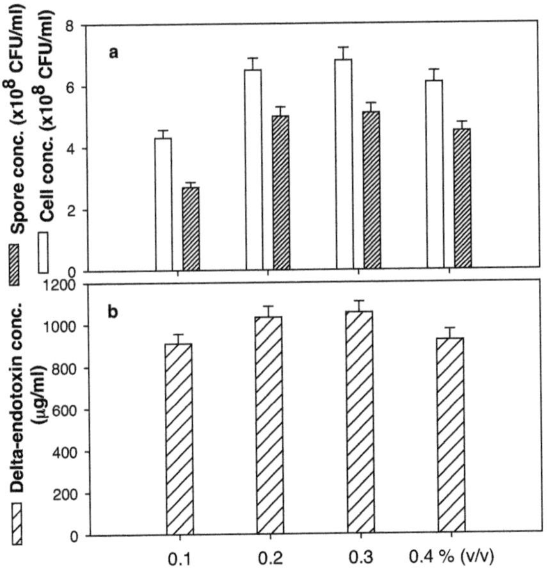

Figure 16. Effects of different concentrations (0.1; 0.2; 0.3; and 0.4 %, v/v) of Tween 80 fortified into SIWS (SIW + cornstarch at 1.25%, w/v) on (a) cell and spore concentration and (b) delta-endotoxin concentration

2.3.3.2. Bioreactor experiment

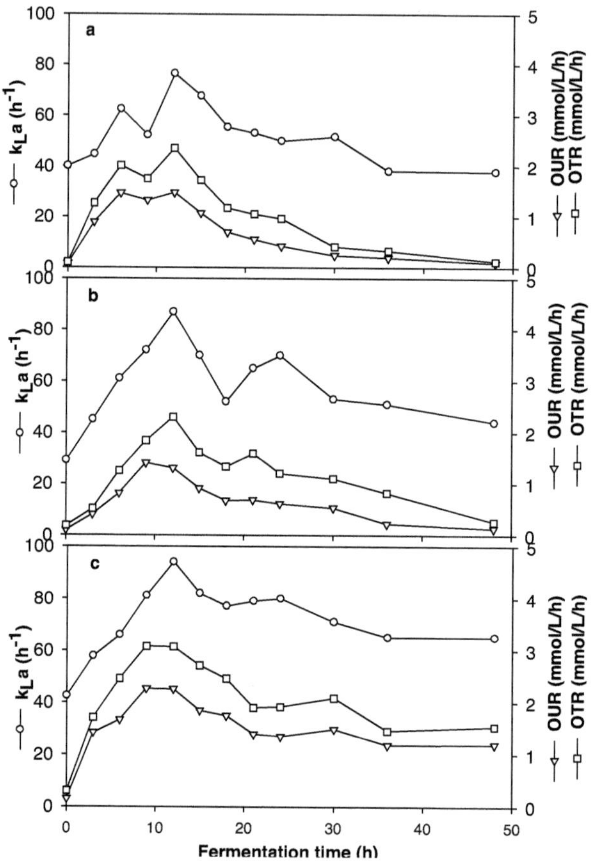

Figure 17. Fermentation parameters kLa, OUR, OTR of (a) Synthetic medium; (b) SIW and (c) SIWST (SIW + cornstarch + Tween 80)

Fermentation of Btk using SIW without any fortification and SIW fortified with both cornstarch (1.25 %, w/v) and Tween 80 (0.2 %, v/v) (designated as SIWST) as raw materials was conducted in 15 L bioreactor. Further, the results were also compared with that of using synthetic medium (semi-synthetic medium based on soy meal) and SIW (without any fortification) which recorded as control.

2.3.3.2.1. Fermentation parameters

It is known that oxygen transfer is one of the most important factors in an aerobic fermentation process (Stanbury et al. 2003). Therefore, profiles of fermentation parameters such as volumetric oxygen transfer coefficient (K_La), oxygen uptake rate (OUR), oxygen transfer rate (OTR) for three different media (synthetic medium, SIW and SIWST) were determined and presented in Figure 17. The maximum values of these parameters are presented in Table 14. Maximum values of OUR, OTR and K_La of SIWST were highest in three media (Table 14) which demonstrated that the presence of Tween 80 in SIWST enhanced oxygen transfer during the fermentation process. The results are in agreement with other research reports who also claimed improvement in oxygen transfer (K_La, OTR and OUR) efficiency with fortification of Tween 80 in various fermentation media (Brar et al. 2005).

Tableau 14. Maximum values of fermentation process parameters in different media

Parameters	Synthetic medium	SIW	SIWST
Max. K_La (h^{-1}) **	76.4 h^{-1} (12h)*	87.1 h^{-1} (12h)	94.0 h^{-1} (12h)
Max OTR ($mmolO_2\ lh^{-1}$) **	2.35	2.3	3.7
Max. OUR ($mmolO_2\ lh^{-1}$) **	1.47	1.4	2.3
Max. specific growth rate ($\mu max\ h^{-1}$) **	0.38	0.43	0.43
Max. total cell count ($x\ 10^8$ CFU ml^{-1}) ***	6.0 ± 0.42^b (15h)	5.70 ± 0.37^b (12h)	11.0 ± 0.62^a (18h)

Max spore count (x 10^8 CFU ml^{-1}) ***	3.4 ± 0.23 [a] (48h)	1.78 ± 0.12 [b] (24h)	6.0 ± 0.41 [a] (48h)
Total cell count at 48h (x 10^8 CFU ml^{-1}) ***	3.8 ± 0.26 [a]	2.6 ± 0.18 [b]	6.8 ± 0.40 [a]
Spore count at 48h (x 10^8 CFU ml^{-1}) ***	3.4 ± 0.24 [a]	1.70 ± 0.11 [b]	6.0 ± 0.4 [a]
*Max. sporulation at 48h (%)***	89.5 ± 6.3 [a]	64.4 ± 5.20 [b]	88.2 ± 6.1 [a]
Max delta-endotoxin concentration (µg/ml) at 48h ***	1020 ± 51 [b]	515.7 ± 25.8 [c]	1327.0 ± 66.4 [a]
Max Tx of fermented broth (x 10^6 SBU/ml) ***	11.7 ± 0.85 [c]	15.6 ± 1.1 [b]	18.1 ± 1.2 [a]
SpTx-spore (SBU/1000 spores) of fermented broth	34.4	91.8	30.2
SpTx-toxin (x 10^3 SBU/µg delta-endotoxin) of fermented broth	11.5	30.3	13.6
Spore count (x10^8 CFU/ml) in suspended pellet ***	32.3 ± 2.3 [b]	16.2 ± 1.1 [c]	57.0 ± 4.0 [a]
Delta-endotoxin concentration (µg/ml) in suspended pellet ****	10200 ± 510 [b]	5157 ± 258 [c]	13270 ± 664 [a]
Tx of suspended pellet (x 10^6 SBU/ml) ***	22.7 ± 1.6 [b]	21.3 ± 1.5 [b]	26.7 ± 1.9 [a]
SpTx-spore (SBU/1000 spores) of fermented broth	7.0	13.2	4.7
SpTx-toxin (x 10^3 SBU/µg delta-endotoxin) of fermented broth	2.2	4.1	2.0

* Digits in parenthesis represent the time of fermentation at which maximum values of different parameters occurred

** The presented values are the mean values obtained from two separate experiments conducted for each fermentation condition

*** The values are the mean of three determinations of two separate experiments conducted for each fermentation condition. The presented values are the mean ± SD. Different letters (stand near each value) within the same row indicate the least significant differences among these values determined by one-factor analysis of variance (Fisher LSD test, $p \leq 0.05$)

**** Delta-endotoxin concentration in suspended pellet derived from delta-endotoxin concentration in fermented broth which was concentrated 10 times

2.3.3.2.2. Cell, spore and delta-endotoxin concentrations during fermentation

Profiles of total cell count (TC), vaiable spore (VS) during growth of Btk in different media are depicted in Figure 18a and the maximum values of growth parameters (maximum cell and spore counts, maximum specific growth rate, and maximum % sporulation) are illustrated in Table 14. Use of SIWST gave considerable higher values of TC, VS as compared to SIW and synthetic medium (Table 14). Maximum spore concentration was obtained at 24 h for SIW and 48 h for synthetic medium and SIWST (Table 14). This is mainly due to the fact that available nutrients in SIW are lower than other cases (synthetic medium and SIWST). The results further confirmed the role of cornstarch (as carbon source) and Tween 80 (as carbon and surface active agent) fortification in SIW in augmenting TC, VS as demonstrated in shake flask experiments. Further, the concentration of cell, spore and delta-endotoxin observed in bioreactor were higher than those observed in shake flask experiments (Table 14 and Figure 16) and is attributed to the control of pH and DO in bioreactor.

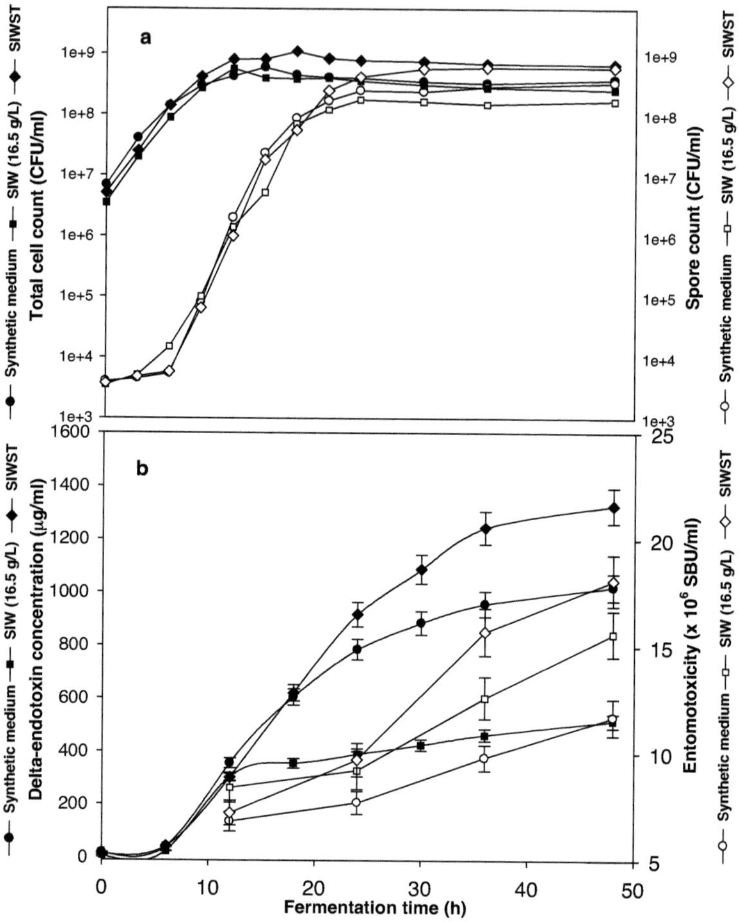

Figure 18. Profiles of (a) total cell count and spore count and (b) delta-endotoxin and entomotoxicity during fermentation using synthetic medium, SIW and SIWST (SIW + cornstarch + Tween 80)

Profiles of delta-endotoxin concentration produced by Btk in 3 different media during the fermentation are presented in Figure 18b. Delta-endotoxin was present at the beginning of fermentation (low concentration) and increased during fermentation until the end of the process. Except in case of

SIW where the increase in delta-endotoxin concentration from 24 h to 48 h was very slow compared to other media. It is known that delta-endotoxin production is concomitant with the sporulation (to produce sporulated cells) of Btk and the lysis of sporulated cells release heat resistant spore and delta-endotoxins into the medium (Schnepf et al., 1998). As mentioned above, maximum concentration of spores in SIW occurred at 24 h and that could be the reason for high concentration of delta-endotoxin at 24 h of SIW fermentation. Comparatively, very slow increase in delta-endotoxin concentration between 24 to 48 h could be due to the accumulation of delta-endotoxin caused by the lysis process of the residual sporulated cells in the medium. Delta-endotoxin concentration obtained in SIWST was highest in three medium (Table 14) which also confirmed the important role of supplementation of cornstarch and Tween 80 in SIW.

2.3.3.2.3. Entomotoxicity (Tx) and specific entomotoxicity of fermented broth

Profiles of Tx values during the fermentation of Btk in three media are presented in Figure 18b. Tx value of fermented broth (containing delta-endotoxin, spores, viable cells, enzymes, cell debris, intracellular components, vegetative insecticidal proteins -VIPs, among others) was measured from 12h and it was found to increase until the end of fermentation (48h). Tx values at 48h of fermentation followed the order: SIWST > SIW > synthetic medium (Table 14). The differences were statistically significant ($p \leq 0.05$). Generally, in each medium Tx of fermented broth increased with delta-endotoxin concentration (Figure 18b). However, the increase in Tx value was not proportional to the delta-endotoxin concentration. For example, in case of SIWST, maximum delta-endotoxin concentration (1327.0 μg/ml) was more than 2 times higher than that of SIW (515.7 μg/ml) but the Tx value of SIWST fermented broth was only 16 % higher than Tx of fermented broth of SIW or in case of synthetic medium, maximum delta-endotoxin concentration (1020 μg/ml) was 2 times higher than that of SIW (515.7 μg/ml) but the Tx value of fermented broth of synthetic medium was less than that of SIW (Table 14). The reason behind these variations is that Tx of fermented broth is affected by many factors other than delta-endotoxin and spore concentration. Firstly, Btk might have produced vegetative insecticidal proteins (Vips – a group of proteins produced during vegetative growth) which could act independently as biopesticide to kill target insect pests (Estruch et al. 1996; Donovan et al. 2001). Vips production depends mainly on the type of components available in the fermentation media. Thus, Btk while growing in SIW would have produced Vip3A during the fermentation which would have acted independently and synergistically with spores and delta-endotoxin (in fermented broth) to provide higher Tx value. The second factor that might affect the Tx value is the production of Zwitttermicin A (an antibiotic) during fermentation. In fact, this antibiotic which produced by *Bacillus cereus* UW85 can act synergistically with Btk in commercial biopesticide against Gypsy moth (Broderick et al. 2000). Like Vips, Zwitttermicin A synthesis also depends on the medium components (Milner et al. 1995). It is known that Btk can produce Zwittermicin A (Stabb et al. 1994). Therefore, it is likely that Zwittermicin A could be produced at different concentrations during growth of Btk in different media and thus contributing towards the Tx values. Another possible reason could be the nonenzymatic glycosylation of crystal proteins (containing delta-endotoxins components) occurred at different levels depending on medium that might cause changes in their activities against tareget insect pest (Bhattacharya et al. 1993). It is known that (a) the nonenzymatic glycosylation occurs preferentially on lysine side chains of protein at alkaline pHs; (b) the *Bacillus thuringiensis* protoxin (delta-endotoxins) is activated in the larval gut by trypsin-like enzymes and trypsin only cleaves peptide bonds on the carboxyl side of lysine or arginine residues. Therefore, the glycosylation of the lysine amino group could alter the proteolytic cleavage pattern of these protoxins and finally the entomotoxicity of activated toxins against target insect pest might change (Bhattacharya et al. 1993).

Specific Tx values in term of delta-endotoxin – SpTx-toxin (SBU/μg delta-endotoxin) and specific Tx values in term of spore – SpTx-spore (SBU/1000 spore) of fermented broths of three media were calculated and presented in Table 14. These values varied largely for different media and should be attributed to many factors such as vegetative insecticidal proteins (Vips), antibiotic (Zwittermicin A), phospholipases etc. These factors might act synergistically against target insect pest (Donovan et al. 2001; Stabb et al. 1994). It is interesting to note that the highest values of SpTx-spore and SpTx-toxin occurred in case of SIW in which the lowest nutrients are available as compared to synthetic medium and SIWST. The exact reason is not known, however, it is supposed that Btk cultivated in SIW might produce other synergistic factors at higher concentration as compared to other medium. Moreover, it should be emphasized here that the quality of produced delta-endotoxin is very important. Liu and Bajpai (1995) also demonstrated that quality of toxin produced by Bt can affect the Tx value against specific insect pest. The same fact was also recognized in our previous research (Vu et al. 2008). Thus, it is difficult to predict Tx value based only on spore or delta-endotoxin concentration in a fermented broth.

2.3.3.2.4. Tx and specific Tx of suspended pellets (spore and delta-endotoxin complex)

It has been specified that the main component for killing the target insect pest are delta-endotoxins which act specifically against insect pest and the spores may play a synergistic role in killing some target insect pests (Schnepf et al. 1998). Thus, for practical application entomocidal potency or Tx of Bt based biopesticides is normally assessed based on the complex of spore and delta-endotoxin which is harvested by centrifugation of the fermented broth (Dulmage et al. 1970; Brar et al. 2006). The Tx value (SBU/L) against spruce budworm larvae of suspended pellets obtained from three different fermented media are summarized in Table 14. The difference in Tx values of suspended pellets of synthetic medium and SIW is statistically insignificant (Fisher LSD test, $p \leq 0.05$), but Tx value of suspended pellet of SIWST is significantly higher than Tx values of the other two medium (Fisher LSD test, $p \leq 0.05$).

In our previous research, where optimm total solids (TS) concentration of SIW (30 g/L) was used as fermentation medium, Tx value of suspended pellet was 25.8 ±1.8 x 10^9 SBU/L (or 25.8 ± 1.8 x10^6 SBU/ml) (Vu et al. 2009). In this research, the Tx value of suspended pellet using SIWST was 26.7 ± 1.9 x 10^6 SBU/ml (Table 14). These Tx values are not significantly different. Therefore, use of optimized TS concentration of SIW or SIWST as fermentation medium for Btk production leads to similar Tx values and therefore it warrants to examine the economics of the two processes. In fact, use of optimum TS concentration 30 g/L need an extra preparation step such as settling of SIW followed by dispensing the settled solids to a desired concentration (30 g/L) (Vu et al. 2009). This preparation step may be disadvantageous compared to fortification of SIW with cornstarch and Tween 80 (SIWST-this research), comparatively an easy step. However, in case of SIWST the supplementation of cornstarch (1.25 %, w/v) and supplementation of Tween 80 (0.2 %, v/v) may increase the fermentation cost. Further, cornstarch and Tween could be fortified to SIW with TS 30 g/L and that could probably further enhance cell, spore, delta-endotoxin conacnetraion and Tx value. This study is in progress in our laboratory.

Specific Tx values in terms of delta-endotoxin – SpTx-toxin (SBU/µg delta-endotoxin) and specific Tx values in term of spore – SpTx-spore (SBU/1000 spore) of suspended pellets of three fermented media were also calculated and presented in Table 14. These values also varied depending on the type of fermented broth used to obtain the suspended pellet. The highest values of SpTx-spore and SpTx-toxin also occurred in case of SIW where the lowest concentrations of spore and delta-endotoxins were present. The results further demonstrated that Tx value against the target insect pest could not be predicted based on only spore or delta-endotoxin or even both spore and delta-endotoxin concentration. And therefore, bioassay against the target insect pest must be conducted to estimate the Tx of Bt based biopesticides.

2.3.4. Conclusion

The following conclusions can be drawn based on the foregoing research:

- Starch industry wastewater (SIW) contains sufficient enough nitrogen to support growth as well as the synthesis of δ-endotoxin of Btk; however, lacks in carbon source to achieve maximum growth, synthesis of δ-endotoxin and entomotoxicity value.
- Fortification of Tween 80 at 0.2 %, v/v and cornstarch at 1.25 %, w/v to SIW in shake flask experiments significantly enhanced cell, spore, delta-endotoxin concentrations and entomotoxicity as compared to the control. Fermentor results also confirmed the highest entomotoxicity value in cornstarch and Tween 80 fortified medium.
- Entomotoxicity of the fermented broth and of the suspended pellet increased by 16.0% and 24.9 %, respectively, with fortification of Tween and cornstarch.

2.3.5. Acknowledgements

The authors are sincerely thankful to the Natural Sciences and Engineering Research Council of Canada (Grant A4984, Canada Research Chair) for financial support. Scicerely thanks to Ms. Jyothi Benzawada for checking English. The views and opinions expressed in this article are strictly those of authors.

2.3.6. References

1. Aiba S, Humphrey AE, Millis NF (1973) Biochemical Engineering, 2nd ed. Academic Press, New York.
2. Avignone-Rossa C, Arcas J, Mignone C (1992) *Bacillus thuringiensis*, sporulation and δ-endotoxin production in oxygen limited and nonlimited cultures. World J Microbiol Biotechnol 8:301-304.

3. Bhattacharya M, Plantz BA, Swanson-Kobler JD, Nickerson KW (1993) Nonenzymatic glycosylation of Lepidopteran-active *Bacillus thuringiensis* protein crystals. Appl Environ Microbiol 59:2666-2672.
4. Brar SK, Verma M, Barnabe S, Tyagi RD, Valero JR, Surampalli RY (2005) Impact of Tween 80 during Bacillus thuringiensis fermentation of wastewater sludges. Process Biochem 40:2695-705.
5. Brar SK, Verma M, Tyagi RD, Valéro JR, Surampalli RY (2006) Efficient centrifugal recovery of *Bacillus thuringiensis* biopesticides from fermented wastewater and wastewater sludge.Water Res 40:1310-1320.
6. Beegle CC (1990) Bioassay methods for quantification of *Bacillus thuringiensis* δ-endotoxin. In: Hickle LA and Fitch WL (eds) Analytical Chemistry of *Bacillus thuringiensis*. American Chemical Society, NewYork, pp. 14–21.
7. Bradford MM (1976) A rapid and sensitive method for the quantitation of microgram quantitites of protein utilizing the principle of protein-dye binding. Analytical Biochem 72:248-254.
8. Broderick N., Goodman RM, Raffa KF, Handelsman J (2000) Synergy between Zwittermicin A and *Bacillus thuringiensis* subsp. *kurstaki* against Gypsy Moth (Lepidoptera: Lymantriidae). Environ Entomol 29:101-107.
9. Dulmage HT, Correa JA, Martinez AJ (1970) Coprecipitation with lactose as a means of recovering the spore-crystal complex of *Bacillus thuringiensis*. J Invertebr Pathol 15:15-20.
10. Dulmage HT, Boening OP, Rehnborg CS, Hansen GD (1971) A proposed standardized bioassay for formulations of *Bacillus thuringiensis* based on the international unit. J Invertebr Pathol 18:240-245.
11. Donovan WP, Donovan JC, Engleman JT (2001) Gene knockout demonstrates hat *vip3A* contributes to the pathogenesis of *Bacillus*

thuringiensis toward *Agrotis ipsilon* and *Spodoptera frugiperda*. J Invertebr Pathol 78:45-51.

12. Estruch JJ, Warren GW, Mullins MA, Nye GJ, Craig JA, Koziel MG (1996) Vip3A, a novel *Bacillus thuringiensis* vegetative insecticidal protein with a wide spectrum of activities against lepidopteran insects. Proc Nat Acad Sci USA 93:5389-5394.
13. Leblanc ME (2003) Effets des différentes stratégies et prétraitements des biosolides municipaux sur la croissance, la sporulation et l'entomotoxicité de *Bacillus thuringiensis* var. *kurstaki*. MSc. Thesis, INRS-Université du Québec.
14. Liu WM, Bajpai RK (1995) A Modified Growth Medium for *Bacillus thuringiensis*. Biotechnol Prog 11: 589-591.
15. Milner JL, Raffel SJ, Lethbridge BJ, Handelsman J (1995) Culture conditions that influence accumulation of zwittermicin A by *Bacillus cereus* UW85. Appl Microbiol Biotechnol 43:685-691.
16. Prescott LM , Harley JP , Klein DA (2002) Microbiology 5th Edition. New York, NY McGraw-Hill. Page 192.
17. Scherrer P, Luthy P, Trumpi B (1973). Production of δ-Endotoxin by *Bacillus thuringiensis* as a function of glucose concentrations. Appl Microbiol 25:644-646.
18. Schnepf E, Crickmore N, van Rie J, Lereclus D, Baum J, Feitelson J, Zeigler DR, Dean DH (1998) *Bacillus thuringiensis* and its pesticidal crystal proteins. Microbiol Mol Biol Rev 62:775-806.
19. Singh A, Van Hamme JD, Ward OW (2007) Surfactants in microbiology and biotechnology: Part 2. Application aspects. Biotechnol Adv 25:99-121.
20. Stabb EV, Jaconson LM, Handelsman J (1994) Zwittermicin A-producing strains of Bacillus cereus from diverse soils. Appl Environ Microbiol 60:4404-4412.

21. Stanbury PF, Whitaker A, Hall SJ (2003) Principles of fermentation technology, 2nd edition. Butterworth-Heinemann, Burlington, MA.
22. APHA, AWWA, WPCF (1998) Standard methods for examination of water and wastewaters, 20th ed. American Public Health Association, Washington, DC.
23. Vu KD, Tyagi RD, Valéro JR, Surampalli RY. (2009a) Impact of different pH control agents on biopesticidal activity of *Bacillus thuringiensis* during the fermentation of starch industry wastewater. Bioprocess Biosyst Eng 32: 511-519.
24. Vu KD, Tyagi RD, Brar SK, Valéro JR, Surampalli RY (2009b) Starch industry wastewater for production of biopesticides- ramifications of solids concentrations. Environ Technol 30:393-405.
25. Yezza A, Tyagi RD, Valéro JR, Surampalli RY (2006) Bioconversion of industrial wastewater and wastewater sludge into *Bacillus thuringiensis* based biopesticides in pilot fermentor. Bioresource Technol 97:1850-1857.
26. Zouari N, Dhouib A, Ellouz R, Jaoua S (1998) Nutritional requirements of a strain of *Bacillus thuringiensis* subsp. *kurstaki* and use of gruel hydrolysate, for the formulation of a new medium for d-endotoxin production. Appl Biochem Biotechnol 69:41-52.
27. Zouari N, Jaoua S (1999) The effect of complex carbon and nitrogen, salt, Tween-80 and acetate on delta-endotoxin production by a *Bacillus thuringiensis* subsp. *kurstaki*. J Industrial Microbiol Biotechnol 23:497-502.

Partie 2.4. Production de *Bacillus thuringiensis* (Btk HD-1) par bioréaction en Fed-batch en utilisant les eaux usées d'industrie d'amidon comme milieux de fermentation

Résumé

Des biopesticides à base de Btk HD-1 ont été produits par fermentation soit en batch soit en Fed-batch en employant des eaux usées d'industrie d'amidon comme seuls substrats de culture. Le mode de culture Fed-batch a été conduit en se basant sur le taux d'oxygène dissous (DO) comme paramètre de contrôle rétroactif et le mode manuel d'intermittence de l'alimentation en éléments nutritifs présents dans les SIW dans le milieu de fermentation. Il a été démontré que la production en Fed-batch avec deux alimentations intermittentes (à 10 et 20 h) au cours d'une fermentation de 72 heures a donné le maximum au point de vue de la concentration en delta-endotoxines (1672,6 µg/ml) et de la valeur du pouvoir insecticide (Tx :18,5 x 10^6 SBU/ml) dans le bouillon fermenté et qui sont significativement plus élevées que les maximum en delta-endotoxines (511,0 µg/ml) et Tx (15,8 x 10^6 SBU/ml) obtenus dans la fermentation en batch ou d'autres en Fed-batch avec une ou trois alimentations. La fermentation en Fed-batch avec trois alimentations intermittentes (à 10, 20 h et 34 h) pourrait améliorer de façon significative la concentration en cellules, mais il y a eu apparition de Btk non sporogènes, ce qui a provoqué une baisse de la production de spores, de delta-endotoxines et de la valeur de la Tx dans le bouillon fermenté. Les valeurs de Tx de suspensions des culots des fermentations en batch et en Fed-batch variaient selon l'ordre suivant: fermentation en Fed-batch avec deux alimentations (27.4 x 10^6 SBU/ml) > fermentation en Fed-batch avec un apport (22.7 x 10^6 SBU/ml) ≥ fermentation en Fed-batch avec trois ajouts (21.8 x 10^6 SBU/ml) ≥ fermentation en batch (21.0 x 10^6 SBU/ml).

Batch and Fed-batch fermentation of *Bacillus thuringiensis* using starch industry wastewater as fermentation and fed substrate

Abstract

Bacillus thuringiensis var. *kurstaki* biopesticide was produced in batch and fed batch fermentation modes using starch industry wastewater as sole substrate. Fed batch fermentation with 2 intermittent feeds (at 10 and 20 h) during the fermentation of 72 h gave the maximum delta-endotoxin concentration (1672.6 mg/L) and entomotoxicity (Tx) (18.5 x 10^6 SBU/ml) in fermented broth which were significantly higher than maximum delta-endotoxin concentration (511.0 mg/L) and Tx (15.8 x 10^6 SBU/ml) obtained in batch process. However, fed batch fermentation with 3 intermittent feeds (at 10, 20 and 34 h) of the fermentation resulted in the formation of asporogenous variant (Spo-) from 36 h to the end of fermentation (72 h) which resulted in a significant decrease in spore and delta-endotoxin concentration and finally the Tx value. Tx of suspended pellets (27.4 x 10^6 SBU/ml) obtained in fed-batch fermentation with 2 feed was the highest value as compared to other cases.

Keywords: *Bacillus thuringiensis*; starch industry wastewater; fed batch culture; entomotoxicity; delta-endotoxin

2.4.1. Introduction

Entomtoxicity (Tx) or biopesticidal activity of *Bacillus thuringiensis* (Bt)-based biopesticides against the target insect pest is usually closely related to delta-endotoxin and spore concentration in the final products [1-4]. In addition, spore and delta-endotoxin production may be significantly affected by the nutritional composition of the medium [2-6]. Therefore, the production of Bt delta-endotoxin and spore can often be improved by changing or optimizing the culture parameters [1].

On an industrial scale, Bt has been produced mostly by batch process [6], however, the spore concentration and delta-endotoxin production obtained in batch process could be improved [5]. Bt has also been produced in semi-continuous and continuous culture. The latter procedure is limited by the spontaneous development of asporogeneuos mutants (the Spo^- mutants formed translucent colonies while the wild-type colonies were opaque) during long cultivation time [7]. On the other hand, total cell retention culture (TCRT) (using a bioreactor incorporating with a ceramic membrane filter to improve spore concentration) and intermittent fed-batch culture (IFBC) have been applied to *Bacillus thuringiensis* var *kurstaki* production using semi-synthetic medium and a great success in which TCRT and IFBC could achieve spore concentration of 1.6 x 10^{10} and $1.25x10^{10}$ (CFU/ml), respectively [7-8]. However, in these researches, delta-endotoxin concentration and entomotoxicity (Tx) of Bt *subsp. kurstaki* biopesticide of both culture methods (TCRT and IFBC) against target insect pest were not reported.

In recent years, the research focused on utilisation of wastewater sludge as sole substrate for fermentation of Bt and reported very promising results on Tx value as compared to that of using semi-synthetic medium [9]. Application of fed-batch mode of fermentation enhanced the Tx value by 38% over the batch process [9]. Starch industry wastewater (SIW) was investigated as sole substrate for Bt based biopesticides production and the results demonstrated high Tx potential which could be used for practical application [3-4]. Therefore, in this study, fed-batch process was explored to enhance growth, spore formation, delta-endotoxin production and the Tx against spruce budworm larvae (Lepidoptera: *Choristoreuna fumiferena*) employing SIW as sole substrate.

2.4.2. Materials and Methods

2.4.2.1. Bacterial strain and inoculum preparation

Bacterial strain: *Bacillus thuringiensis* var. *kurstaki* HD-1 (ATCC 33679) (Btk) was used in this study. The Btk was subcultured and streaked on tryptic soya agar (TSA) plates, incubated for 48 hour at 30 ± 1°C and then preserved at 4 °C for future use [9].

Inoculum preparation: All the media used for inoculum preparation were adjusted to pH 7 before autoclaving. A loopful of Btk from TSA plate was used to inoculate a 500-ml Erlenmeyer flask containing 100 ml of sterilised tryptic soya broth (TSB) medium. The flask was incubated on a rotary shaker at 250 revolutions per min (rpm) and 30 °C for 8–12 h. A 2% (v/v) inoculum from this flask was then used to inoculate 500-ml Erlenmeyer flasks containing 100 ml of sterilised starch industry wastewater (SIW) and incubating for 8-12 h. Then, the actively growing cells from these flasks were used as an inoculum (pre-culture) for the production of Btk-based biopesticide in bioreactor.

2.4.2.2. Bt production medium

SIW was collected from ADM-Ogilvie (Candiac, Québec, Canada) and used as a raw material for Btk growth. Total solids (TS), total volatile solids (TVS), suspended solids (SS) and volatile suspended solids (VSS), pH, total carbon, total Kjeldahl nitrogen (N_t), ammonia nitrogen ($N–NH_3$) and total phosphorus (P_t) were determined according to Standard Methods [10]. Physico-chemical characteristics of SIW are presented in Table 15.

Further, SIW was also settled and separated to obtain liquid and solids fractions. Then, solids and liquid fraction were mixed at different ratios to obtain different TS concentrations such as 20, 30, 40 and 50 g/L TS [4].

Tableau 15. Characteristics of starch industry wastewater (SIW)

Parameter	**Concentration*(g/L)**
Total solids [TS]	15.10 ± 0.70
Total volatile solids [TVS]	10.62 ± 0.50
Suspended solids [SS]	5.15 ± 0.20
Volatile suspended solids [VSS]	5.03 ± 0.20
pH	3.60 ± 0.20
	Concentration* (g/kg TS)
Ct	397± 7
N_t	28 ± 1
$N\text{-}NH_3$	0.70 ± 0.01
P_t	7.0 ± 0.2

* The presented values are the mean ± standard deviation (SD).

2.4.2.3. Fermentation procedure

Fermentation experiments were conducted in bioreactors (15 L total capacity) equipped with accessories and automatic control systems for dissolved oxygen, pH, antifoam, impeller speed, aeration rate and temperature. SIW with total solids (TS) of 15 g/L was filled in the bioreactor (10 L for batch mode and 8 L for fed batch mode) and polypropylene glycol (PPG, Sigma- Canada) (0.1% v/v) solution was added to control the foam during sterilization. The fermenters were sterilized *in situ* at 121 °C for 20 min. After sterilization, fermenters were cooled to 30 ◦C. The fermenters were then inoculated (2% v/v inoculum) with actively growing cells of the pre-culture. The agitation speed (300–500 rpm) and aeration rate (2-4 litres per minute or LPM) were varied in order to keep the dissolved oxygen (DO) values above 30% of saturation, which ensured the oxygen concentration above the critical level [11].The pH was controlled at 7.0 ± 0.1 using either 4N NaOH or 4 N H_2SO_4 through computer-controlled peristaltic pumps.

2.4.2.4. Fed batch operation

Preliminary research on fed-batch mode was conducted on shake-flask fermentation for 72 h. The feed substrate was SIW with different total solids (20, 30, 40 and 50 g/L) and the feed was added at 10 h interval (based on DO recorded during batch fermentation in bioreactor). The preliminary results indicated that feed substrate with total solids (TS) of 30 g/L gave better results in terms of spore, delta-endotoxin concentration and Tx of fermented broth at the end of fermentation when compared with other TS (20, 40 and 50 g/L) of feed substrate (data not show). Therefore, feed substrate with TS of 30 g/L was used in all experiments in the fed-batch mode.

In this research, fed-batch strategy was conducted based on dissolved oxygen (DO) as feedback control parameter and using manual intermittent feeding mode to feed the nutrients (present in SIW) into the fermentor. The fermentor was fed one, two or three times and the feeding time was determined based on DO measurement during the fermentation process. Fermentation was started in batch mode with a working volume of 8 L. Following inoculation, the DO concentration began to drop until the end of exponential growth. After that, DO started increasing. When DO tended to stabilize, a fresh pre-sterilized SIW with TS of 30 g/L (2 L) was added. The DO again decreased to a minimum followed by an increase. Once the DO reached near maximum, 3L of fermenter contents was withdrawn and replaced by the same volume of sterilized fresh SIW with TS of 30 g/L. The DO value again decreased followed by an increase and reached near maximum, 3L of fermenter contents was withdrawn and replaced by the same volume of sterilized fresh SIW with TS of 30 g/L. Then, the fermentation process was continued in batch mode until 72 h. The samples were drawn from the fermenters at regular intervals to determine cell and spore counts, delta-endotoxin concentration and Tx value.

At the end of fermentation, fermented broths from different experiments (batch and fed-batch) were adjusted to pH 4.5 and centrifuged at 9000 g for 30 min [3]. The obtained pellets (spore and delta-endotoxin complex) were re-suspended with supernatant to obtain one-tenth volume of the original fermented broth. The suspended pellets were subjected to bioassay.

2.4.2.5. Analysis of parameters

2.4.2.5.1. Determination of K_La, OUR and OTR

The volumetric oxygen transfer coefficient (K_La) measurement was based on a dynamic method [12-13]. This technique consists of interrupting the air input. Afterwards, the aeration is re-established. The decrease and the increase in DO concentration were recorded. K_La was determined from the mass balance on DO just after sampling of fermentation broth. During batch fermentation, the mass balance of the DO concentrations could be written as:

$dC_L/dt = OTR - OUR$

where: OTR: oxygen transfer rate from gas phase to liquid phase, $OTR = k_La(C* - C_L)$; OUR: oxygen uptake rate : $OUR = QO_2X$; K_La: volumetric oxygen transfer coefficient; $C*$: saturated oxygen concentration; C_L: dissolved oxygen concentration in the medium; QO_2 : specific oxygen uptake rate; and X: cell concentration

Oxygen concentration in the fermentation broth was converted from % air saturation to mmol O_2/Las follows: the DO electrode was calibrated in medium at 30°C and then transferred to air saturated distilled water at known temperature and ambient pressure. This reading was used; with the known saturation concentration of oxygen in distilled water (0.07559mmol/L) (100%), to estimate the saturation concentration of oxygen in the cultivation media at 30°C.

2.4.2.5.2. Estimation of total cell count (TC) and spore count (SC)

The TC and *SC* were performed by counting colonies grown on TSA medium [9]. For all counts, the average of at least triplicate plates was used for each tested dilution. For enumeration, 30–300 colonies were enumerated per plate. The results were expressed as colony forming units per mL (CFU/ml).

2.4.2.5.3. Estimation of delta-endotoxin concentration produced during the fermentation

Delta-endotoxin concentration was determined based on the solubilization of insecticidal crystal proteins in alkaline condition [2]: 1 ml of samples collected during fermentation was centrifuged at 10 000 g for 10 min. at 4°C. The pellet containing a mixture of spores, insecticidal crystal proteins, cell debris and residual suspended solids was used to estimate the concentration of alkaline solubilised insecticidal crystal proteins (delta-endotoxin): These pellets were washed three times with 1 ml of 0.14 M NaCI - 0.0 1 % Triton X- 100 solution. The washing helped in eliminating the soluble proteins and proteases which might have adhered on the pellets and could affect the integrity of the crystal protein. The insecticidal crystal proteins in the pellet were solubilized with 0.05N NaOH (pH 12.5) for three hours at 30°C with stirring. The suspension was centrifuged at 10 000 g for 10 min. at 4°C and the pellet, containing spores, cell debris and residual suspended solids was discarded. The supernatant, containing the alkaline solubilised insecticidal crystal proteins was used for determination of delta-endotoxin concentration by Bradford method [14] using bovine serum albumin as standard protein. The values presented were the average results (± SD) of triplicates of two independent experiments.

2.4.2.5.4. Bioassay technique

The potential Tx of Btk HD-1 was estimated by bioassay against third instar larvae of spruce budworm (Lepidoptera : *Choristoreuna fumiferena*) following the diet incorporation method [15-16]. The commercial preparation 76B FORAY (Valent Bioscience Corporation Libertyville, IL) was used as a reference standard to analyse the Tx. The detail procedure of bioassay technique was presented in Yezza et al., (2005) [9]. The Tx was evaluated by comparing the mortality percentages of dilutions of the samples (fermented broth or suspended pellet) with same dilutions of the standard. In our research, Tx was expressed as relative spruce budworm units (SBU/L). On comparison of Tx of Bt-fermented samples, it was found that SBU reported in our study was 20–25% higher than IU [9]. The presented values are the mean of three determination of two independent experiments ± SD.

2.4.3. Results and Discussion

2.4.3.1. Batch fermentation of Bacillus thuringiensis using SIW as raw material (Run # 1)

SIW is known to contain a dissolved fraction (consist of soluble components such as reducing sugars, soluble protein, minerals, etc.) and a solids fraction (consist of insoluble components such as residual starch, fibres, nitrogen, etc.). It was demonstrated that ratio between dissolved fraction and solids fraction in SIW higher than 0.45 was more favourable for Btk growth to obtain maximum cell concentration in a shorter fermentation time [4]. Therefore, in this research TS of 15 g/L was used in batch fermentation or in fed batch fermentation. Batch fermentation of Btk using 10 L SIW (TS of 15 g/) was conducted in bioreactor for 48h. Profiles of batch fermentation (Run # 1) parameters such as dissolved oxygen (DO), agitation, aeration, volumetric mass transfer coefficient (K_La), oxygen uptake rate (OUR), oxygen transfer rate (OTR) and profiles of total cell count, spore count, delta-endotoxin concentration and Tx of fermented broth during the process are illustrated in Figure 19.

After inoculation, DO level was decreased from the beginning to 10 h of fermentation. To keep the DO above critical level, agitation speed was increased (Figure 19. The decrease in DO level was due to the increase in

OUR associated with active growth of bacterial population. The highest

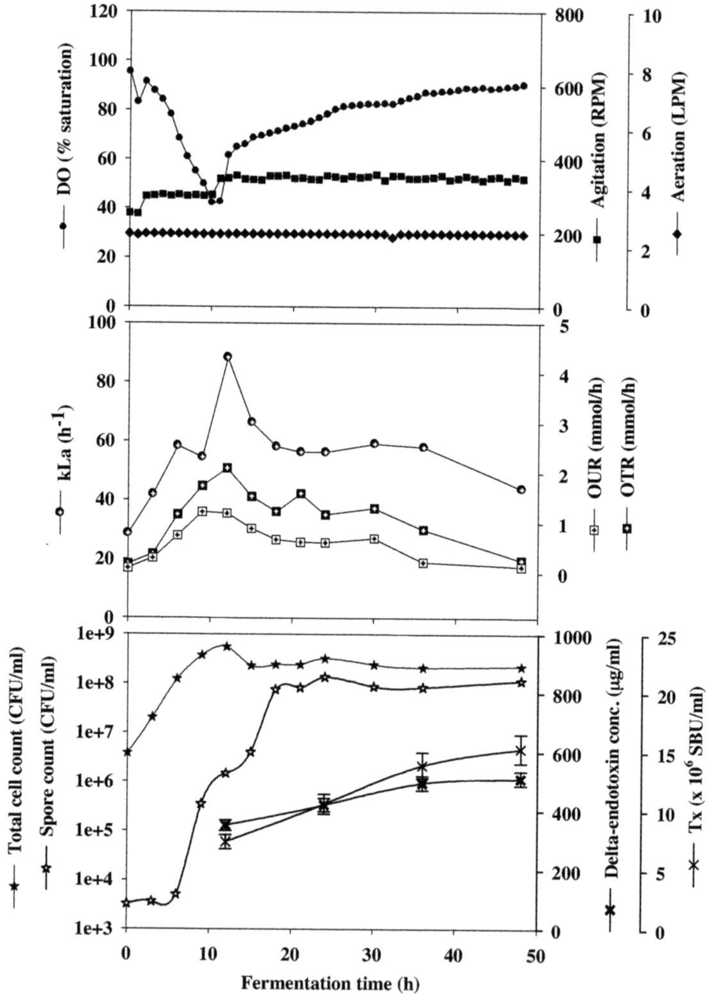

Figure 19. Run # 1, Batch fermentation of Bt

values of OUR, OTR and K_La occurred at 12 h (Figure 19). When the substrate in the medium was exhausted, the bacteria became substrate limited consequently decreased their oxygen demand, resulting in an increase in the DO level and decrease in OUR, OTR and K_La value from 12 h until the end of fermentation (Figure 19).

Exponential phase ended at approximately 12 h when substrate was exhausted or DO started increasing and the total cell count (CFU/ml) attained maximum value (Table 16). The decline in cell concentration from 12 h might be due to limitation of nutrients present in the medium causing cell death. Spore concentration increased significantly from 6 to 18 h and reached maximum value at 24h. Delta-endotoxin concentration also increased significantly from 12 to 30 h and reached maximum value at 48 h (Figure 19 and Table 16); however, the values of delta-endotoxin concentration at 36 and 48 h were not significantly different (Figure 19). It is generally understood that delta-endotoxin is synthesized concomitantly with spore formation and both are released at the same time in the medium during cell lysis [17]; however, increase in spore count did not match increase in delta-endotoxin concentration (Figure 19). Spores attained maximum value in 24 hrs, whereas delta-endotoxin kept increasing and reached maximum at 48 h (at the end of the fermentation). It is possible that at 24 h of fermentation, fermented medium consisted of vegetative cells, sporulated cells, free spores and free crystal proteins (one crystal protein contains about 1×10^6 to 2×10^6 delta-endotoxin molecules) and other components [17]. The sporulated cells (unlysed) continued to lyse (from 24 to 48 h of fermentation) and released crystal proteins (consequently it kept increasing) and spores in the medium. Tx increased significantly from 12 to 36 h of fermentation and attained maximum value at 48h (Figure 19 and Table 16), however, the Tx values at 36 h and 48 h are not statistically significantly different (Figure 19). It should be noted here that maximum spore concentration occurred at 24 h which did not match with the maximum Tx at 48h (Figure 19).

Tableau 16. Maximum values of cell, spore, delta-endotoxin concentrations and Tx

Run number	**# 1**	**# 2**	**# 3**	**# 4**
Fermentation mode	Batch	Fed-batch	Fed-batch	Fed-batch
Feeding number	0	1	2	3
Intermittent feeding time (h)	no	10	10 and 20	10, 20 and 34
Max. cell count (x10^8 CFU/ml) [A]	5.60 ± 0.33 [d]	8.50 ± 0.51 [c]	25.00 ± 0.15 [b]	38.00 ± 0.23 [a]
	(12 h)	(18 h)	(36 h)	(48 h)
Max. spore count (x 10^8 CFU/ml) [A]	1.40 ± 0.08 [c]	2.90 ± 0.17 [b]	13.00 ± 0.08 [a]	2.80 ± 0.19 [b]
	(24 h)	(30 h)	(48 h)	(36 h)
Increase in spore count (times) [B]	1.00	2.15	9.63	2.07
Max. delta-endotoxin concentration (μg/ml) [A]	511 ± 24 [d]	1,018 ± 50.1 [c]	1,672.6 ± 83.9 [a]	1,134.6 ± 56.7 [b]
	(48 h)	(48 h)	(48 h)	(36 h)
Increase in delta-endotoxin concentration (times) [B]	1	2	3.3	2.2
Max. Tx of fermented broth (x10^6 SBU/ml) [A]	15.8 ± 1.2 [b]	17.4 ± 1.4 [ab]	18.5 ± 1.5 [a]	16.5 ± 1.3 [ab]
	(48 h)	(48 h)	(48 h)	(36 h)
Increase in Tx of fermented broth (%) [B]	0	10.1	17.1	4.4

SpTx (x10^3 SBU/µg delta-endotoxin) in fermented broth	30.9	17.0	11.0	14.5
SpTx-spore (SBU/1000 spore) in fermented broth	139.8	62.1	14.2	58.9
Max. Tx of suspended pellet (x10^6 SBU/ml)	21.0 ± 1.7 [b] (48h)	22.7 ± 1.8 [b] (48h)	27.4 ± 1.9 [a] (48 h)	21.8 ± 1.7 [b] (36 h)
Increase in Tx of suspended pellet (%) [B]	0	8.1	30.5	3.8
Spore concentration in suspended pellet (x 10^8 CFU/ml) [A]	10.7 ± 0.75 (48h)	26.6 ± 1.9 (48h)	124.0 ± 8.6 (48h)	24.7 ± 1.7 (48h)
Delta-endotoxin concentration in suspended pellet (µg/ml) [C]	5,110 ± 240 (48 h)	10,180 ± 501 (48 h)	16,726 ± 839 (48 h)	11,346 ± 567 (36 h)
SpTx (x10^3 SBU/µg delta-endotoxin) in suspended pellet	4.1	2.2	1.6	1.9
SpTx-spore (SBU/1000 spore) in suspended pellet	19.6	8.8	2.2	8.8

[A] The results are the mean of three replications. The letter stand near the number was the indicator of the Fisher Least Significantly Difference ($P < 0.05$, Statistica 7.0).

[B] The increase (%) was calculated based on batch fermentation.

[C] Values derived from fermented broth (concentrated 10 times).

It is probably due to the fact that cell lysis did not completely occur and thus a substantial amount of the delta-endotoxin was still cell bound and was not available to participate in the killing mechanism of the insects. As the delta-endotoxin was released in the medium or free delta-endotoxin concentration in the medium increased, the Tx value increased. Thus, Tx value is mainly dependent on free delta-endotoxin concentration in the medium and not only the spore concentration (Figure 19).

Spore and delta-endotoxin concentrations of suspended pellets were 10.7 x 10^8 (CFU/ml) and 5110 (μg/ml), respectively. Tx values of fermented broth and suspended pellet (complex of spore and delta-endotoxin- concentrated 10 times from the fermented broth samples) of batch fermentation 15.8 x 10^6 and 21.0 x 10^6 (SBU/ml), respectively (Table 16). To further enhance the cell, spore, delta-endotoxin concentrations as well as Tx values of fermented broth and suspended pellet, fed-batch strategy was developed (1 feed, 2 feeds or 3 feeds) to achieve maximum values of parameters mentioned above.

2.4.3.2. Fed-batch fermentation of Bacillus thuringiensis using SIW as feed-substrate

2.4.3.2.1. Fed batch with 1 feed at 10 h (Run # 2)

Profiles of DO, agitation, aeration, K_La, OUR, OTR, total cell count, spore count, delta-endotoxin concentration and Tx during fed-batch fermentation (Run # 2) are illustrated in Figure 20.

Fermentation was started in batch mode with working volume of 8 L. The concentration of DO decreased sharply and reached the lowest value at 8 h, followed by a slight increase in DO from 8 to 10 h. At this time (10 h), the substrate (2 L of SIW with TS of 30 g/L) was fed into the fermentor which caused a little decrease in DO from 10 to 11 h. DO again increased from 11 to 19 h, decreased from 19 h to 26 h and finally increased again until the end of fermentation (Figure 20).

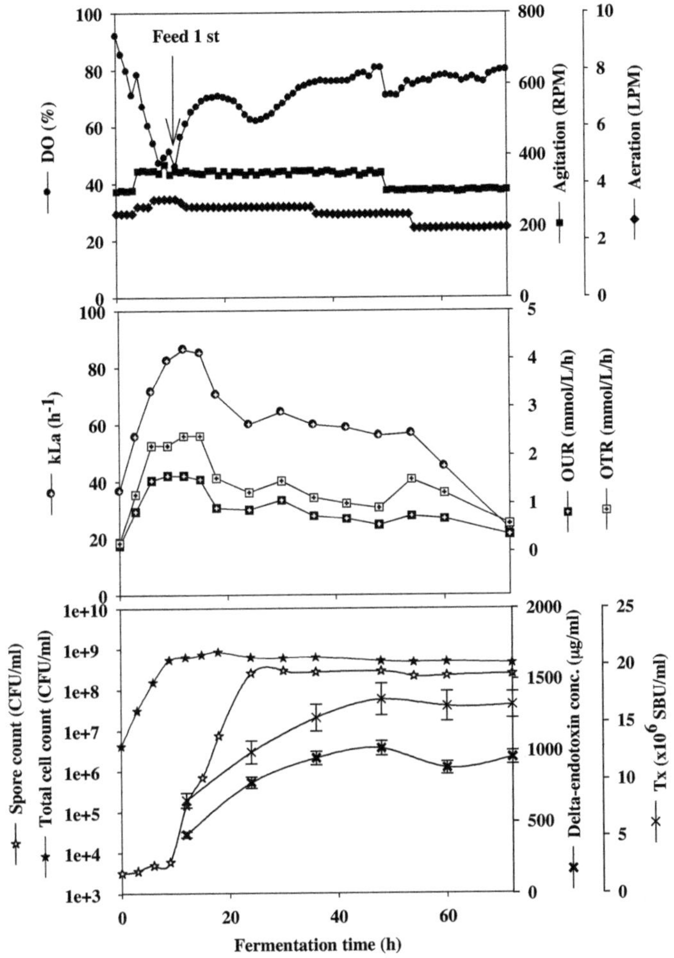

Figure 20. Run # 2, Fed-batch fermentation of Bt with 1 feed at 10h

Values of K_La, OUR and OTR increased from 0h to 9 h, were almost stable from 9 to 15 h, followed by a decrease until the end of fermentation (Figure 20). It is known that K_La, OTR and OUR values are important parameters to monitor [12-13]. The high and stable values of these parameters from 9 to

15 h confirmed that Bt population in the medium during this period was actively growing (high values of OUR).

Total cell concentration increased exponentially from 0 to 9 h followed by a slow increase from 12 to 18h to reach maximum value, followed by a slight decrease until the end of fermentation (Figure 20 and Table 16). Spore concentration increased significantly from 9 to 30 h and attained maxium during this period (Figure 20 and Table 16). Delta-endotoxin concentration and Tx value also increased significantly from 12 to 48h (to reach maximum values) (Figure 20 and Table 16). From 48 to 72 h, spore and delta-endotoxin concentrations and Tx of fermented broth were rather stable. This fact demonstrated that prolonged fermentation time in this case was not necessary.

The interesting point is that employing one time feed (at 10 h); the spore and delta-endotoxin concentrations was 2.15 times and 2.0 times higher than spore and delta-endotoxin concentrations obtained in the batch process (Run # 1), respectively (Table 16). However, maximum Tx value with one feed was slightly higher (not statistically significant) as compared to Tx value of the batch fermentation (Run # 1) (Table 16). It is well known that Tx value is mostly based on the concentration of delta-endotoxin and spores [18]. Accordingly, Tx of fermented broth of one feed must have been higher than Tx of batch fermentation significantly, however, the difference was not significant. The possible reason for this fact is that there might have other synergistic factors (Vegetative Insecticial Proteins-Vips, Zwittermicin A-an antibiotic, etc.) which could be produced by Btk at different concentrations in fermented broth of batch fermentation and fed-batch fermentation with one feed. In fact, Zwitttermicin is known to be produced and depended on the medium components [19]. It is also known that Btk HD-1 (the strain used in this research) can produce Zwittermicin A [20]. For example, this antibiotic which produced by *Bacillus cereus* UW85 can act synergistically with Btk in commercial biopesticide against Gypsy moth [21]. Production of non-identified antibiotic by Btk active against several human pathogens has been also detected in our laboratory (un-published results). Therefore, it is likely that Zwittermicin A could be produced in larger concentration during growth of Btk in batch fermentation than in fed-batch fermentation with one feed. And therefore, the synergistic action of Zwittermicin A with spore-delta-endotoxin caused the Tx value of batch fermentation to be almost similar to that of fed-batch fermentation with one feed (even with significant higher spore and delta-endotoxin concentration).

In case of suspended pellet, the maximum Tx value in fed batch fermentation with one feed was also slightly higher as compared to the batch process, however, the difference was also not statistically significant (Table 16). The results further confirmed that using fed batch mode with one feed (at 10 h) could enhance delta-endotoxin and spore concentration significantly, however, it could not significantly increase Tx in the fermented broth as well as the suspended pellet. Therefore, further experiment with 2 intermittent feed was conducted and the results are presented below.

2.4.3.2.2. Fed batch with 2 feeds at 10 and 20 h (Run # 3)

Profiles of DO, agitation, aeration, K_La, OUR, OTR, total cell count, spore count, delta-endotoxin concentration and Tx value obtained during two feeds fed-batch process (Run # 3) are illustrated in Figure 21.

The first feed (at 10 h) was the same as in Run # 2, 2 L of SIW with TS of 30 g/L was added into the fermentor which caused a little decrease in DO from 10 to 11 h. The DO level increased from 11 to 20 h, at this time 3 L of fermented broth was withdrawn and then 3 L of fresh feed (sterilised SIW with TS of 30 g/L) was added into the bioreactor (the second feed). The 2nd feed caused a significant decrease in DO level from 20 h to 24 h followed by an increase of DO until 36 h. The DO decreased again between 36 h to 45 h and finally increased again from 45 h to the end of fermentation (Figure 21). The decrease of DO between 36 to 45 h should be due to the usage of the residual complex substrates (starch, glutens, etc.) present in SIW by Btk [22-23].

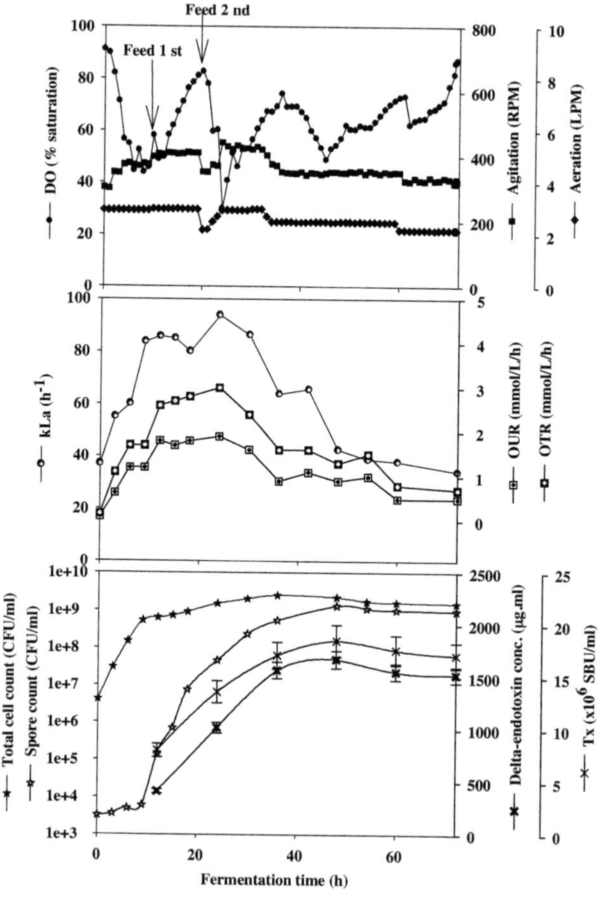

Figure 21. Run # 3, Fed batch fermentation of Bt with 2 feeds at 10h and 20h

Values of K_La, OUR and OTR increased from 0h to 9 h, relatively stable from 9 to 24 h and finally a decrease from 24 h until the end of fermentation (Figure 21). As discussed above, the stable values of these parameters during

9 to 24 h confirmed that Bt population in this period was actively growing (high values of OUR).

Total cell concentration increased exponentially from 0 to 9 h followed by a slow increase from 12 to 36 h (also attained maximum value in this period) followed by a slight decrease until the end of fermentation (Figure 20). Spore concentration increased significantly from 9 to 48 h (Figure 20). Delta-endotoxin concentration and Tx of the fermented broth also increased significantly from 12 to 48 (Figure 21). From 48 to 72 h, spore and delta-endotoxin concentrations and Tx values were almost stable or slightly decreased. The slight decrease might be due to the hydrolysis of delta-endotoxin (main component that contributes towards Tx value) by intracellular proteases which were also released during the lysis of sporulated cells [24]. The results also found that the prolonged time of fermentation in fed-batch process with 2 feeds is also not necessary.

The results demonstrated that two intermittent feeds (at 10 and 20 h) did provide significantly higher concentration of spores (9.63 times) and delta-endotoxin (3.3 times) compared to batch fermentation (Run # 1) (Table 16). Tx values of fermented broth and suspended pellet of fed batch fermentation with 2 feeds increased by 17.1 % and 30.5 % as compared to those obtained in the batch process, respectively (Table 16). Here, the interesting question arise that how much Tx is higher compared to our earlier batch results at 30g/LTS in SIW? [4]. In fact, there is no significant difference. The exact reasons for this fact are not known, however, it is probably due to the fact that in batch culture, as said before, Btk may produce more soluble components (Vip, Zwittermicin A, etc.) than that in fed-batch culture and therefore, these component can act synergistically with delta-endotoxin and spore againt spruce budworm and finally caused the Tx value in batch culture almost the same as in fed-batch culture. The second possible reason is that it is not known if the quality of delta-endotoxin produced in batch or fed-batch mode is the same. That was the reason why even delta-endotoxin concentration was estimated in our research, the bioassay was always conducted parallel to be sure that the obtained results are authentic.

These results further confirmed that the fed batch mode could enhance the synthesis of delta-endotoxin, spore and finally the Tx of the fermented broth. However, in this case only 2 feeds were applied and it was not known if increasing number of feeds to 3 could further enhance cell, spore and delta-endotoxin concentration as well as Tx values of the fermented broth and the suspended pellet. Therefore, 3 intermitten feeds fed-batch process was conducted and the results were presented in the following section.

2.4.3.2.3. Fed batch with 3 feeds at 10, 20 and 34 h (Run # 4)

The profiles DO, agitation, aeration, K_La, OUR, OTR and profiles of total cell count, spore count, delta-endotoxin concentration and Tx value obtained in 3 feed fed-batch process (Run # 4) are illustrated in Figure 22.

The first and second feeds (at 10 and 20 h) in this experiment were the same as in earlier experiment (Run # 3), after the second feeding, DO level decreased from 20 h to 24 h and then increased from 24 h to 34 h, at this time 3 L of fermented broth was withdrawn and 3 L of feed (sterilised SIW with TS 30 g/L) (the third feed) was added to the fermentor. This feed caused a significant decrease in DO level from 34 h to 39 h, DO level increased again from 39 h until the end of fermentation (Figure 22).

Generally, values of K_La, OUR and OTR were high during the period of 9 h to 48 and finally decreased from 48 h until the end of fermentation (Figure 21). The high values of these parameters during the period of 9 to 48 h confirmed that Bt population in the medium in this period was actively growing (high values of OUR).

Total cell concentration increased exponentially from 0 to 9 h and then slightly increased from 12 to 48 h in which maximum value occurred, followed by a slight decrease (insignificant) until the end of fermentation

(Figure 22 and Table 16). The concentration of delta-endotoxin and Tx value

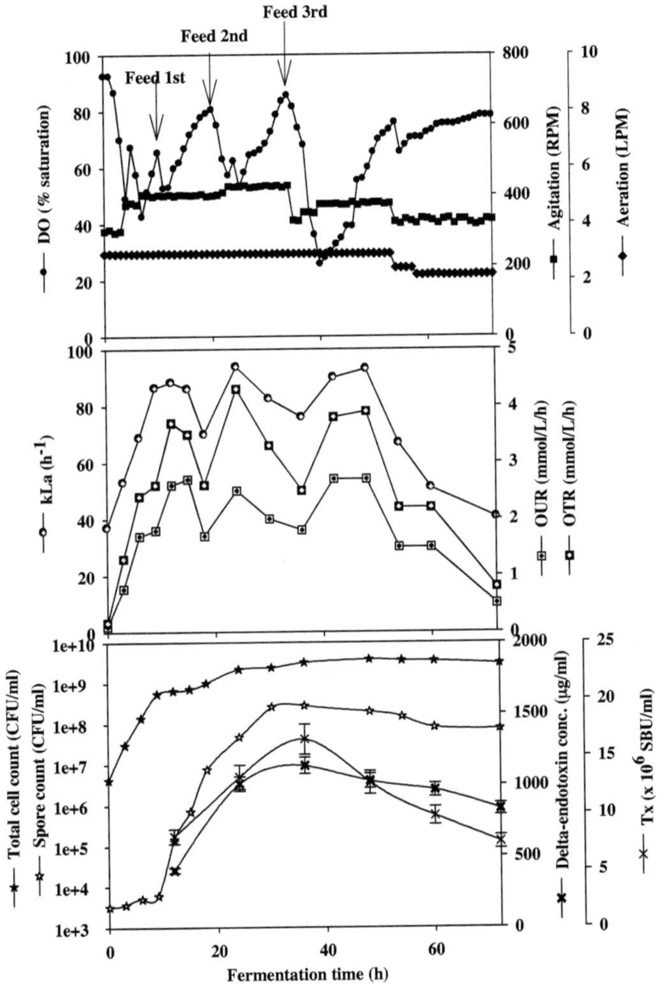

Figure 22. Run # 4, Fed batch fermentation of Bt with 3 feeds at 10h, 20h and 34h

decreased after 36 h of the process. During this period, there was an appearance of asporogenous variants (Spo-) (these mutant spores produced translucent colonies as compared to the wide-type of Btk was opaque) of Btk in fermented broth from 36 h to the end of fermentation. The exact reason for the appearance of Spo^- variants is not known. In fact, it is well known that continuous culture of Btk or Bt. *galleriae* caused the appearance of asporogenous variants (Spo^-) [7, 25]. According to Sachidanandham and Jayaraman (1993) [25], the absence of critical nutritional factors in the culture medium, which could be essential for the maintenance of genetic elements governing sporulation, might have also led to the loss of sporulation. Moreover, in the continuous culture of *Bacillus sphaericus* 2362, the appearance of asporogenous variants was also observed and the entomotoxicity of fermented product against mosquito was also significantly decreased as compared to the wide-type variant (spo^+) [26]. These authors also found growth of the mutants population in a relatively short period (within 72 h) and they suggested that the mutants had a selective advantage over other variants within the population [26].

The obtained results in this case demonstrated that three intermittent feeds (at 10, 20 and 30 h) could achieve significantly higher cell concentration (38.0 x 10^8 CFU/ml), but attained lower spore and delta-endotoxin concentration as compared to values obtained in fed-batch fermentation with two feeds (Table 2). Hence, it is possible that fed-batch culture with many feeds or continuous culture might not be suitable for the production of Btk based biopesticides when using SIW as sole raw material.

2.4.3.3. Specific Tx of fermented broth and suspended pellet

2.4.3.3.1. Specific Tx of fermented broth

Tx of fermented broth increased with delta-endotoxin concentration (Table16). However, the increase in Tx value was not proportional to the delta-endotoxin concentration. For example, in case of fed batch with one feed, delta-endotoxin concentration (1018.0 mg/L) was 2 times higher than that of batch fermentation (511.0 mg/L) but the Tx value was not significantly higher (discussed above) than Tx of batch fermented broth

(Table 16). The same fact was also noticed with spore concentration. For example, in case fed batch fermentation with 2 feeds, spore concentration obtained was 9.6 times higher than that of batch fermentation but the Tx value was only 17.1% higher.

Specific Tx values in term of delta-endotoxin – SpTx-toxin (SBU/μg delta-endotoxin) and specific Tx values in term of spore – SpTx-spore (SBU/1000 spore) of fermented broths (from batch and fed-batch processes) were calculated and presented in Table 16. These values varied largely and attributed to many factors such as delta-endotoxin, spore, vegetative cells and many soluble substances such as vegetative insecticidal proteins (Vips), antibiotic (Zwittermicin A) etc. These factors might act synergistically against target insect pest [21, 27]. Moreover, it should be emphasized here that the quality of produced delta-endotoxin is very important. Liu and Bajpai, (1995) [28] also demonstrated that quality of toxin produced by Bt can affect the Tx value against specific insect pest. The same fact was also recognized in our previous research [3]. Thus, it is difficult to predict Tx value based only on spore or delta-endotoxin concentration in fermented broth.

2.4.3.3.2. Specific Tx of suspended pellet

Specific Tx values in term of delta-endotoxin – SpTx-toxin (SBU/μg delta-endotoxin) and specific Tx values in term of spore – SpTx-spore (SBU/1000 spore) of suspended pellet (from batch and fed-batch processes) were also calculated and presented in table 16. These values also varied depending on the fermented broth used to obtain suspended pellet. The results further demonstrated that Tx value against the target insect pest could not be predicted based on only spore or delta-endotoxin or even both spore and delta-endotoxin concentration.

2.4.4. Conclusions

Based on the foregoing study employing batch and fed-batch fermentation of *Bacillus thuringiensis* using starch industry wastewater as sole substrate, the following conclusions could be drawn:

- On-line DO profile during cultivation of *Bacillus thuringiensis* could be used as a convenient indicator for identifying feed time.
- Intermittent fed-batch fermentation with 2 feeds (at 10 h and 20 h) could be a best strategy for enhancing Btk cell, spore and delta-endotoxin production and the final entomotoxicity of the fermented broth and entomotoxicity of the suspended pellet.
- Intermittent fed-batch fermentation with 3 feeds (at 10 h, 20 h and 30 h) could enhance significantly the cell concentration. However, there was the appearance of asporogenous variants of Btk which caused a decrease in spore and delta-endotoxin production and the entomotoxicity of fermented broth and that of suspended pellet.
- Entomotoxicity of fermented broth didn't increase proportionally with the increase in spore or delta-endotoxin concentration.
- Values of specific entomotoxicity (SBU/μg delta-endotoxin, SBU/1000 spores) varied with number of feeds. The entomotoxicity value was not predictable based on only spore or delta-endotoxin concentration in the fermented broth or in the suspended pellet.

2.4.5. Acknowledgements

The authors are sincerely thankful to the Natural Sciences and Engineering Research Council of Canada (Grant A4984, Canada Research Chair) for financial support. Sincere thanks to Mr. Mathieu Drouin and Mr. Jean-Philippe Chenel for helping us in sampling starch industry wastewater. The views and opinions expressed in this article are strictly those of authors.

2.4.6. References

1. Dulmage HT (1971) Production of δ-Endotoxin by eighteen isolates of Bacillus thuringiensis, Serotype 3, in 3 fermentation media. J Invertebr Pathol 18: 353-358
2. Zouari N, Jaoua S (1999) The effect of complex carbon and nitrogen, salt, Tween-80 and acetate on delta-endotoxin production by a *Bacillus thuringiensis* subsp *kurstaki*. J Ind Microbiol Biotechnol 23:497–502
3. Vu KD, Tyagi RD, Valéro JR, Surampalli RY. (2008) Impact of different pH control agents on biopesticidal activity of *Bacillus thuringiensis* during the fermentation of starch industry wastewater. Bioprocess Biosyst Eng. DOI 10.1007/s00449-008-0271-z
4. Vu KD, Tyagi RD, Brar SK, Valéro JR, Surampalli RY (2009) Starch industry wastewater for production of biopesticides- ramifications of solids concentrations. Environ Technol 30: 393-405
5. Arcas J, Yantorno OM, Ertola RJ (1987) Effect of high concentration of nutrients on *Bacillus thuringiensis* cultures. Biotechnol Lett 9: 105-110
6. Pearson D, Ward OP (1988) Effect of culture conditions on growth and sporulation of *Bacillus thuringiensis* subsp. *israelensis* and development of media for production of the protein crystal endotoxin. Biotechnol Lett 10:451–456
7. Kang BC, Lee SY, Chang HN (1993) Production of *Bacillus thuringiensis* spores in total cell retention culture and two-strages continuous culture using an interal ceramic filter system. Biotechnol Bioeng 42: 1107-1112.
8. Kang BC, Lee SY, Chang HN (1992) Enhanced spore production of *Bacillus thuringiensis* by fed-batch culture. Biotechnol Lett 14: 721-726.
9. Yezza A, Tyagi RD, Valéro JR, Surampalli RY (2005) Production of *Bacillus thuringiensis*-based biopesticides in batch and fed batch

cultures using wastewater sludge as a raw material. J Chem Technol Biotechnol 80:502–510

10. APHA, AWWA, WPCF (1998) Standard methods for examination of water and wastewaters, 20th ed.. In Clesceri LS, Greenberg AE, Eaton AD (ed). American Public Health Association, Washington, DC
11. Avignone-Rossa C, Arcas J and Mignone C (1992) *Bacillus thuringiensis* sporulation and δ-endotoxin production in oxygen limited and non-limited cultures. World J Microbiol Biotechnol 8:301–304
12. Aiba S, Humphrey AE, and Millis NF (1973) Biochemical Engineering, 2nd edn. Academic Press, New York
13. Stanbury PF, Whitaker A, Hall SJ (2003) Principles of fermentation technology, 2nd edition. Butterworth-Heinemann, Burlington, MA
14. Bradford MM (1976) A rapid and sensitive method for the quantitation of microgram quantitites of protein utilizing the principle of protein-dye binding. Analytical Biochem 72: 248-254
15. Dulmage HT, Boening OP, Rehnborg CS, Hansen GD (1971) A proposed standardized bioassay for formulations of *Bacillus thuringiensis* based on the international unit. J Invertebr Pathol 18: 240-245
16. Beegle CC (1990) Bioassay methods for quantification of *Bacillus thuringiensis* δ- endotoxin. In Hickle LA, Fitch WL (ed) Analytical chemistry of *Bacillus thuringiensis*. American Chemical Society, Washington, DC. pp 14-21.
17. Agaisse H, Lereclus D (1995) How does *Bacillus thuringiensis* produce so much insecticidal crystal protein? J Bacteriol 177: 6027-6032
18. Schnepf E, Crickmore N, van Rie J, Lereclus D, Baum J, Feitelson J, Zeigler DR, Dean DH (1998) *Bacillus thuringiensis* and its pesticidal crystal proteins. Microbiol Mol Biol Rev 62: 775-806

19. Milner JL, Raffel SJ, Lethbridge BJ, Handelsman J (1995) Culture conditions that influence accumulation of zwittermicin A by *Bacillus cereus* UW85. Appl Microbiol Biotechnol 43: 685-691
20. Stabb EV, Jacobsen LM, Handelsman J (1994) Zwittermicin A-producing strains of *Bacillus cereus* from diverse soils. Appl Environ Microbiol 60: 4404-4412
21. Broderick NA, Goodman RM, Raffa KF, Handelsman J (2000) Synergy between Zwittermicin A and *Bacillus thuringiensis* subsp. *kurstaki* against Gypsy Moth (Lepidoptera: Lymantriidae). Environ Entomol 29: 101-107
22. Sklyar V, Epov A, Gladchenko M, Danilovich D, Kalyuzhny S (2003) Combined biologic (anaerobic-aerobic) and chemical treatment of starch industry wastewater. Appl Biochem Biotechnol 109: 253-262
23. Jin B, van Leeuwen HJ, Patel B, Yu Q (1998) Utilisation of starch processing wastewater for production of microbial biomass protein and fungal α-amylase by *Aspergillus oryzae*. Bioresource Technol 66: 201-206
24. Donovan WP, Tan Y, Slaney AC (1997) Cloning of the *nprA* gene for neutral protease A of *Bacillus thuringiensis* and effect of in vivo deletion of *nprA* on insecticidal crystal protein. Appl Environ Microbiol 63: 2311-2317
25. Sachidanandham R, Jayaraman K (1993) Formation of spontaneous asporogenic variants of *Bacillus thuringiensis* subsp, *galleriae* in continuous cultures. Appl Microbiol Biotechnol 40:504-507
26. Idachaba MA, Rogers PL (2001) Production of asporogenous mutants of Bacillus *sphaericus* 2362 in continuous culture. Lett Appl Microbiol 33: 40-44
27. Donovan WP, Donovan JC, Engleman JT (2001) Gene knockout demonstrates hat *vip3A* contributes to the pathogenesis of *Bacillus*

thuringiensis toward *Agrotis ipsilon* and *Spodoptera frugiperda*. J Invertebr Pathol 78: 45–51

28. Liu WM, Bajpai RK (1995) A Modified growth medium for *Bacillus thuringiensis*. Biotechnol Prog 11: 589-591

Partie 2.5. Production induite des chitinases pour améliorer l'entomotoxicité de *Bacillus thuringiensis* (Btk HD-1) en utilisant les eaux usées d'industrie d'amidon comme substrat de fermentation

Résumé

La production induite des chitinases au cours de la bioconversion des SIW en biopesticides à base de Btk HD-1 a été étudiée par fermentation en erlenmeyers et en bioréacteurs. Des SIW ont été fortifiées avec des concentrations différentes (0%, 0,05%, 0,1%, 0,2%, 0,3% p/v) de chitine colloïdale et l'effet de ces additions a été constaté en termes de concentrations en cellules et en spores et des activités des chitinases, des protéases et des amylases et de la valeur de l'entomotoxicité (Tx). À la concentration optimale de chitine colloïdale (0,2% p/v), la valeur de la Tx du bouillon fermenté et celle de suspension du culot ont été améliorées de 12,4 x 10^9 (sans la chitine) à 14,4 x 10^9 SBU/L et de 18,2 x 10^9 (sans la chitine) à 25,1 x 10^9 SBU/L, respectivement. En outre, des expériences ont été effectuées en bioréacteurs de 15 L avec des SIW enrichies de chitine colloïdale à 0,2% p/v (désigné comme SIWC) pour induire la production de chitinases par Btk HD-1 et comparativement à l'emploi de SIW non fortifiées avec de la chitine colloïdale comme contrôle. Il a été constaté que les concentrations en cellules, en spores et en delta-endotoxines et les activités des protéases et des amylases ont été réduites alors que la Tx et l'activité des chitinases ont été augmentées dans le bouillon fermenté des SIWC; l'activité des chitinases atteignant une valeur maximale à 24 h (15 mU/ml) et la Tx de suspension du culot atteignant le niveau le plus élevé (26,7 x 10^9 SBU/L) à 36 h de fermentation dans les SIWC. Dans le contrôle (SIW sans chitine colloïdale), la valeur maximale de la Tx de la suspension du culot (20,5 x 10^9 SBU / L) est atteint après 48 h de fermentation. Un effet synergique des chitinases sur les pouvoirs insecticides des delta-endotoxines, des spores a été observé contre les larves de la tordeuse des bourgeons de l'épinette.

Induced Production of Chitinase to Enhance Entomotoxicity of *Bacillus thuringiensis* Employing Starch Industry Wastewater as a Substrate

Abstract

Induced production of chitinase during bioconversion of starch industry wastewater (SIW) to *Bacillus thuringiensis* var. *kurstaki* HD-1 (Btk) based biopesticides was studied in shake flask as well as in computer controlled fermentors. SIW was fortified with different concentrations (0%; 0.05%; 0.1%; 0.2%, 0.3% w/v) of colloidal chitin and its consequences were ascertained in terms of Btk growth (total cell count and viable spore count), chitinase, protease and amylase activities and entomotoxicity. At optimum concentration of 0.2% w/v colloidal chitin, the entomotoxicity of fermented broth and suspended pellet was enhanced from 12.4 x 10^9 (without chitin) to 14.4 x 10^9 SBU/L and from 18.2 x 10^9 (without chitin) to 25.1 x 10^9 SBU/L, respectively. Further, experiments were conducted for Btk growth in a computer-controlled 15 L bioreactor using SIW as a raw material with (0.2% w/v chitin, to induce chitinase) and without fortification of colloidal chitin. It was found that the total cell count, spore count, delta-endotoxin concentration (alkaline solubilised insecticidal crystal proteins), amylase and protease activities were reduced whereas the entomotoxity and chitinase activity was increased with chitin fortification. The chitinase activity attained a maximum value at 24 h (15mU/ml) and entomotoxicity of suspended pellet reached highest (26.7 x 10^9 SBU/L) at 36 h of fermentation with chitin supplementation of SIW. In control (without chitin), the highest value of entomotoxicity of suspended pellet (20.5 x 10^9 SBU/L) reached at 48 h of fermentation. A quantitative synergistic action of delta-endotoxin concentration, spore concentration and chitinase activity on the entomotoxicity against spruce budworm larvae was observed.

Key words: *Bacillus thuringiensis* var. *kurstaki*, delta-endotoxin concentration, chitinase activity, synergistic action, starch industry wastewater, spruce budworm larvae

2.5.1. Introduction

Starch production from wheat, corn, potato, cassava, etc. generates a huge quantity of wastewater containing high concentration of chemical oxygen d emand (COD) that pose serious problems to environment (Annachhatre and Amatya, 2000; Sklyar et al., 2003). In general, the starch industry wastewater (SIW) can be treated by using upflow anaerobic sludge blanket (UASB) reactor to generate methane (Annachhatre and Amatya, 2000) or in a combined biological and chemical treatment processes (Sklyar et al., 2003). In recent years, another approach that has been followed for management of SIW due to economic reasons is the bioconversion of SIW into value added products such as fungal protein and glucoamylase (Jin et al., 1999), lactic acid (Huang et al., 2003), biocontrol agent (Verma et al., 2006) and Bt-based biopesticides (Yezza et al, 2006).

For effective application of biopesticides or to achieve high kill rate of insects pests in the field, higher entomotoxicity (Tx) in fermented broth of *Bacillus thuringiensis* var. *kurstaki* HD-1 (Btk) is desirable for economic reasons. Btk production employing SIW as raw material showed higher Tx value compared to commercial semi-synthetic soy medium (Yezza et al, 2006). The Tx of Btk based biopesticide against spruce budworm produced using SIW as raw material was substantially enhanced by replacing conventional pH control agents H_2SO_4 with CH_3COOH during fermentation (Vu et al, 2008).

Further, chitinase that hydrolyses chitin (a homopolymer of β-1,4-lined N-acetyl glucosamine), is known to act as a synergistic agent for enhancing the Tx value of biopesticides (Smirnoff, 1973; 1974). It has been suggested that chitinase could increase Tx by perforating the peritrophic membrane barrier in the larval midgut and thus increase the accessibility of the *Bacillus thuringiensis* delta-endotoxin molecule to its receptor on epithelial cell membranes (Regev et al. 1996). Perforation of the peritrophic membrane by chitinase was clearly demonstrated in vivo and in vitro examination (Wiwat et al., 2000; Thamthiankul et al., 2004). These findings confirmed the essential role of chitinase in the hydrolysis of the peritrophic membrane, thus allowing penetration by *Bacillus thuringiensis* spores and crystal toxins into the larval hemolymph.

In recent years, there has been a serge in the researches regarding chitinase producing *Bacillus thuringiensis* strains: (1) selection of chitinase-producing *Bacillus thuringiensis* (Bt) (Rojas-Avelizapa et al., 1999; Barboza-Corona et al., 1999; Liu et al., 2002); (2) cloning and expression of chitinase encoding genes of other microbial chitinase in Bt to enhance the Tx value (Wiwat et al. 1996; Tantimavanich et al., 1997; Lertcanawanichakul and Wiwat, 2000; Sirichotpakorn et al, 2001; Tantimavanich et al., 2004; Thamthiankul et al., 2004); (3) production of endogenous chitinase from Bt as well as cloning, sequencing of chitinase genes or characteristics of chitinases from some Bt strains such as Bt var. *kurstaki* (Wiwat et al., 2000; Arora et al., 2003; Driss et al, 2005), Bt var. *parkitani* (Thamthiankul et al., 2001), Bt var. *kenyae* (Barboza-Corona et al, 2003), Bt var. *israelensis* (Zhong et al, 2003), Bt var. *alesti* (Lin and Xiong, 2004) and Bt var. *sotto* (Zhong et al., 2005). However, these reports mostly used synthetic medium supplemented with colloidal chitin for simultaneous chitinase and biopesticides production by *Bacillus thuringiensis*. Various studies have also reported the use of different wastes as raw materials for Bt based biopesticides and chitinases production such as use of proteo-chitinaceous substrate (milled shrimp waste) as the sole source of ingredients (Rojas-Avelizapa et al., 1999) or using wastewater sludge as a raw material (Brar, 2007). Moreover, bacterial chitinases have been demonstrated to be induced by chitin oligomers and low level of N-acetylglucosamine (Gooday, 1990). Specifically, the strain Btk can produce chitinase when grown on the semi-synthetic medium supplemented with colloidal chitin (Guttmann and Ellar, 2000). Thus, one could conclude from the foregoing researches that supplementation of colloidal chitin into SIW would induce chitinase synthesis in Btk.

It has been well established that Tx of Bt is due to the toxin crystal, Bt cell-spore complex, and vegetative insecticidal protein (VIPs) (Adjalle et al. 2007). In industrial practice, toxin, cells and spores are recovered from Bt fermented broth by centrifugation. A part of toxin spores and cells and all VIPs (soluble) are lost in the centrifuged supernatant. Recently, it was demonstrated that the residual toxin, spores, cells, and VIPs could be recovered from the supernatant employing ultrafiltration technique, which in turn could be mixed with the centrifugate or the concentrated toxin, spores and cells to prepare a high Tx value product (Adjalle et al. 2007). Thus, the chitinase produced during Btk fermentation, which would normally be discarded in the supernatant, could be recovered during ultrafiltration along with residual toxin, spores, cells and soluble VIPs and this could further enhance the Tx value of the final product. Therefore, it is well justified to augment synthesis of chitinase during the Btk fermentation.

Thus, the objective of this research was to investigate the consequences of supplemntation of different colloidal chitin concentration on growth, chitinase production and Tx value of Btk in shake flask as well as in computer controlled fermentors employing SIW as a raw material.

2.5.2. Materials and Methods

2.5.2.1. Bacterial strain and inoculum preparation

Bacterial strain: *Bacillus thuringiensis* var. *kurstaki* HD-1 (ATCC 33679) (Btk) was used in this study. The Btk was subcultured and streaked on tryptic soya agar plates [TSA: 3.0% Tryptic Soya Broth (Difco)+1.5% Bacto-Agar (Difco)], incubated for 48 h at 30± 1°C and then preserved at 4 °C for future use.

Inoculum preparation: All the media used for inoculum preparation were adjusted to pH 7 before autoclaving. A loopful of Btk from TSA plate was used to inoculate a 500-ml Erlenmeyer flask containing 100 ml of sterilised tryptic soya broth (TSB) medium. The flask was incubated on a rotary shaker at 220 revolutions per min (rpm) and 30 °C for 8–12 h. A 2% (v/v) inoculum from this flask was then used to inoculate 500-ml Erlenmeyer flasks containing 100 ml of sterilised SIW which supplemented with colloidal chitin to get the different final concentrations of colloidal chitin: 0%; 0.05%; 0.1%; 0.2% and 0.3% (w/v). These flasks were incubated for 8-12 h and the actively growing cells from these flasks were used as an inoculum (pre-culture) for the production of Btk-based biopesticide in shake flasks or in fermentor (Yezza et al., 2006). This pre-culture step was carried out to adapt Btk in SIW fortified with colloidal chitin.

2.5.2.2. Bt production medium supplemented with colloidal chitin

The SIW was collected from ADM-Ogilvie (Candiac, Québec, Canada) and used as a raw material for Btk growth. SIW was utilized within 1–2 weeks for fermentation because long-term storage even at 4 °C would lead to deterioration.

Total solids (TS), volatile solids (VS), suspended solids (SS) and volatile suspended solids (VSS), pH; total carbon (C_t); total Kjeldahl nitrogen (N_t); ammonia nitrogen ($N–NH_3$); total phosphorus; and metals concentration (Al, Ca, Cd, Cr, Cu, Fe, Mg, Mn, Na, Ni, Pb and Zn) were determined according to Standard Methods (APHA et al., 1998). Physico-chemical characteristics of SIW are presented in Table 17.

Tableau 17. Characteristics of starch industry wastewater (SIW)

Parameter	Concentration* (g/L)
Total solids [TS]	15.05 ± 0.70
Volatile solids [VS]	10.62 ± 0.50
Suspended solids [SS]	5.15 ± 0.20
Volatile suspended solids [VSS]	5.03 ± 0.20
pH	3.60 ± 0.20
	Concentration* (g/kg TS)
Ct	397 ± 7
N_t	28 ± 1
$N\text{-}NH_3$	0.70 ± 0.01
P_t	7.0 ± 0.2
$P\text{-}PO_4^{3-}$	3.00 ± 0.02
Ca	4.00 ± 0.06
K	11.0 ± 0.2
Mg	3.00 ± 0.03
Na	84.0 ± 0.3
S	5.0 ± 0.1
	Concentration* (mg/kg TS)
Al	193.8 ± 6.3
Cd	0.13 ± 0.02
Cr	5.53 ± 0.04
Cu	22.2 ± 0.3
Fe	332.5 ± 14.1
Mn	40.7 ± 0.3
Ni	2.48 ± 0.02
Pb	1.06 ± 0.02
Zn	67.85 ±0.63

* The presented values are the mean ± standard deviation (SD).

2.5.2.3. Shake flask fermentation

Erlenmayer flasks containing 100 ml of sterilised SIW were supplemented with different final colloidal chitin concentrations: 0%; 0.05%; 0.1%; 0.2% and 0.3% (w/v) and inoculated with 2% (v/v) of preculture prepared as above. These flasks were incubated in a shaker-incubator (New Brunswick) for 48 h at 30°C, 220 rpm. Samples (fermented broth) were collected at the end of fermentation process (48h) to determine the total cell count (TC), spore count (SC), chitinase activity (CA) and entomotoxicity (Tx) as described in the following sections. At the end of fermentation (48h), the fermented broths (from all above experiments) were adjusted to pH 4.5 and centrifuged at 9000 *g*, 20 °C for 30 min to obtain pellets (complex of spore and delta-endotoxin) (Brar, 2007). These pellets were re-suspended in corresponding supernatants (of each fermented broth) to obtain the suspended pellets with volume one-tenth of the original volume of fermented broths. These suspended pellets were subjected to bioassay to estimate the Tx value. The presented results were the average results of three determinations of two experiments which were conducted separately. From these results, the optimum chitin concentration that gave the highest chitinase activity and the highest Tx value was chosen and subsequently used in experiments conducted in 15 L fermentor under controlled conditions of pH, temperature and dissolved oxygen (DO) concentration as described below.

2.5.2.4. Fermentation procedure in fermentor 15 L

Fermentation was carried out in a stirred tank 15 L fermentor (working volume: 10 L, Biogenie, Que., Canada) equipped with accessories and programmable logic control (PLC) system for dissolved oxygen (DO), pH, anti-foam, impeller speed, aeration rate and temperature. The software (iFix 3.5, Intellution, USA) allowed automatic set-point control and integration of all parameters via PLC.

Before each sterilisation cycle, the polarographic pH-electrode (Mettler Toledo, USA) was calibrated using buffers of pH 4 and 7 (VWRCanada). The oxygen probe was calibrated to zero (using N_2 degassed water) and 100% (air saturated water). Subsequently, fermentor was charged with SIW (10 L) and polypropylene glycol (PPG, Sigma-Canada) (0.1% v/v) solution as an anti-foam agent. The fermentor with medium was sterilised in situ at 121 °C for 30 min. When the fermentor cooled down to 30 °C, DO probe was recalibrated to zero by sparging N_2 gas and 100% saturation by sparging air at agitation rate of 500 rpm. The fermentor was then inoculated (2% v/v inoculum) aseptically with preculture of Btk in exponential phase (8-12 h). In order to keep the DO above 25% saturation, air flow rate and agitation rate were varied between 0.2-0.3 vvm and 300–350 rpm, respectively. This ensured the critical DO level for Bt above 25% (Avignone-Rossa et al., 1992). The temperature was maintained at 30 °C by circulating water through the fermentor jacket.

Fermentation pH was controlled automatically at 7 ± 0.1 through computer-controlled peristaltic pumps by addition of pH control agents: NaOH 4 M/H_2SO_4 3 M (sodium hydroxide 4 M/sulphuric acid 3 M). Dissolved oxygen and pH were continuously monitored by means of a polarographic dissolved oxygen probe and of a pH sensor (Mettler-Toledo, USA), respectively.

Samples were collected periodically to monitor the changes in total cell count (TC), spore count (SC) and Tx (Tx) as described in the following sections. A part of each sample (20 ml) was centrifuged at 8000 rpm in 20 min and 4°C. The supernatant was used to determine alkaline protease activity (PA), amylase activity (AA) and chitinase activity (CA).

Further, samples (at 30, 36 and 48 h) were adjusted to pH 4.5 and centrifuged at 9000 *g* and 20 °C for 30 min to obtain pellets (complex of spore and delta-endotoxin) (Brar, 2007). These pellets were re-suspended in corresponding supernatants. The volume of suspended pellets with volume was one-tenth of the original volume of fermented broths (concentrated 10 times). These suspended pellets were subjected to bioassay to estimate the Tx.

2.5.2.5. Determination of volumetric oxygen transfer coefficient (K_La), Oxygen Transfer Rate (OTR) and Oxygen Uptake Rate (OUR)

The volumetric oxygen transfer coefficient (k_La) measurement was based on the dynamic method (Aiba et al., 1973). This technique consisted in interrupting the air input. Afterwards, the aeration was re-established. DO concentration was recorded during aeration off and aeration on. k_La was determined from the mass balance on DO just after each sampling of the fermentation broth. Details of k_La determination were described in early report (Yezza et al., 2006).

2.5.2.6. Alkaline protease activity (PA) and amylase activity assay (AA)

PA was determined according to modified Kunitz (1947) method. A part of the supernatant was appropriately diluted with borate buffer pH 8.2 ± 0.1. Alkaline protease activity was assayed by incubating 1 ml enzyme solution with 5 ml casein (1.2%, w/v) for 10 min at 37 °C in a constant-temperature water-bath. One proteolytic activity unit was defined as the amount of enzyme preparation required to liberate 1 μmol (181 μg) tyrosine from casein per minute in pH 8.2 buffer at 37°C. The presented values are the mean of three determinations of two separate experiments.

AA was measured at pH 6, 40 °C according to the method of Miller (Miller, 1959; Zouari and Jaoua, 1999). One enzyme unit was defined as the amount of enzyme required to liberate 1 μg of glucose/min under the experimental conditions. The presented values are the mean of three determinations of two separate experiments.

2.5.2.7. Chitinase activity (CA)assay

Colloidal chitin was prepared according to the method described by Roberts and Selitrennikoff (1988). CA measurement involved colorimetric method (Miller, 1959). CA was determined by measuring the amount of the reducing end group, *N*-acetylglucosamine (NAG), degraded from colloidal chitin at pH 4.0, 40 °C as described by Rojas-Avelizapa et al, (1999). One unit of activity was defined as the amount of enzyme that liberated 1 μmol of NAG per min. Negative control tubes contained all components except substrate, and blanks contained all components except the enzyme. The presented values are the mean of three determinations of two separate experiments.

2.5.2.8. Estimation of delta-endotoxin concentration produced during the fermentation

Delta-endotoxin concentration was determined based on the solubilisation of insecticidal crystal proteins in alkaline condition (Rivera, 1998, Zouari and Jaoua, 1999): 1 ml of samples collected during the fermentation was centrifuged at 10000 $_g$ for 10 min at 4°C. The pellet containing a mixture of spores, insecticidal crystal proteins, cell debris and residual suspended solids was used to estimate the concentration of alkali soluble insecticidal crystal proteins (delta-endotoxin): These pellets were washed three times with 1 ml of 0.14 M NaCI - 0.0 1 % Triton X- 100 solutions. The washing helped in eliminating the soluble proteins and proteases that might be adhered on to the pellet and could affect the integrity of the crystal protein. The insecticidal crystal proteins in the pellet were dissolved with 0.05N NaOH (pH 12.5) for three hours at 30°C with stirring. The suspension was centrifuged at 10000 *g* for 10 min. at 4°C and the pellet, containing spores, cell debris and residual suspended solids was discarded. The supernatant, containing the alkali soluble insecticidal crystal proteins was used to determine the delta-endotoxin concentration by Bradford method (Bradford, 1976) using bovine serum albumin as standard protein. The presented values were the mean of three determination of two separate experiments.

2.5.2.9. Estimation of total cell count (TC) and spore count (SC)

To determine TC and SC, the samples were serially diluted with sterile saline solution (0.85% w/v NaCl). The appropriately diluted samples (0.1mL) were plated on TSA plates and incubated at 30 °C for 24 h to form fully developed colonies. For spore count, the appropriately diluted samples were heated in an oil bath at 80 °C for 10 min and then chilled in ice for 5 min. This heat/cold shock lysed the vegetative cells and liberated those spores already formed in the sporulated cells. The TC and SC were performed by counting colonies grown on TSA medium. For cell and spore counts, the average of at least three replicate plates were used for each tested dilution. For enumeration, 30–300 colonies were enumerated per plate. The results were expressed as colony forming units per mL (CFU/ml).

2.5.2.10. Bioassay technique

Entomotoxicity (Tx) of Btk was estimated by bioassay against third instar larvae of spruce budworm (Lepidoptera : *Choristoreuna fumiferena*) following the diet incorporation method of Dulmage et al., 1971 and Beegle, 1990. The commercial preparation 76B FORAY (Abbott Laboratory, Chicago, IL) was used as a reference standard to analyse the Tx. The detail procedure of bioassay technique was presented in Yezza et al., 2006. The Tx was evaluated by comparing the mortality percentages of dilutions of the samples (suspended pellet as well as whole fermented broth) with same dilutions of the standard. In our research, Tx was expressed as relative spruce budworm units (SBU/L). On comparison of Tx of Bt fermented samples, it was found that SBU reported in our study was 20–25% higher than IU (international units) (Yezza et al., 2006). The presented values are the mean of three determinations of two independent experiments.

2.5.3. Results and discussion

2.5.3.1. Shake flask fermentation

The results of TC, SC, CA and Tx of fermented broth at the end of fermentation (48 h) from shake flask experiments at different colloidal chitin concentration (0%; 0.05%; 0.1%; 0.2% and 0.3% w/v) are presented in Figure 23a. An increase in chitin concentration in SIW (SIWC) resulted in a decrease of TC and SC values, whereas considerably increased CA. The

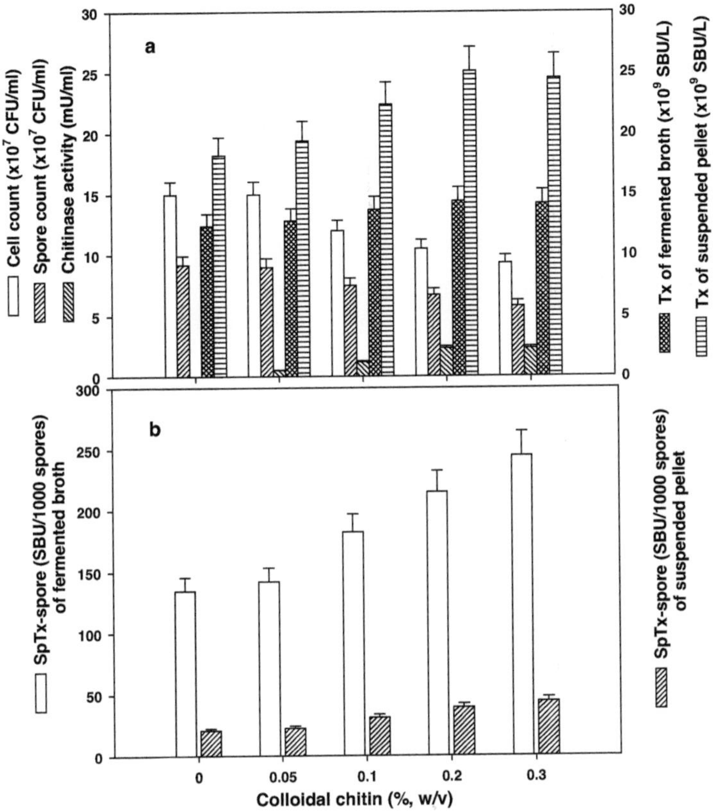

Figure 23. (a)Total cell count, spore count, chitinase activity, Tx of fermented broth, Tx of suspended pellet; and (b) SpTx-spore of fermented broth and suspended pellet at 48h in shake flask supplemented with different colloidal chitin concentrations.

decreased values of TC and SC may be caused by the inhibition due to colloidal chitin supplemented in SIW. Even in spite of decrease in TC and SC counts, an increase in CA resulted in an increase of Tx from 12.4 x10^9 SBU/L in control (without fortification of colloidal chitin) to 14.4 x 10^9 SBU/L at 0.2% w/v chitin concentration, however, the difference was not significant (Figure 23a). Nevertheless, comparing the Tx value taking the similar base (i.e. per 1000 spores, specific entomotoxicity or SpTx-spore), a statistically significant difference could be observed (Figure 23b). Further, in case of suspended pellets (concentrated 10 times), Tx values obtained for SIWC at 0.2 or 0.3% (w/v) chitin concentration were significantly different than those for SIWC at chitin concentration 0.05% (w/v) or SIW (control – without chitin) (Figure 23a). The SpTx-spore values were also significantly different (Figure 1b). The SpTx-spore values were higher at lower spore concentration (fermented broth) than those at higher spore concentration (the concentrated pellet suspension) (Figure 23b) and were in agreement to our earlier published work (Vu et al., 2008). The CA values were almost similar at 0.2 and 0.3 % (w/v) colloidal chitin concentration; however, the Tx value in case of 0.2% (w/v) was slightly higher than 0.3% (w/v) colloidal chitin concentration. Therefore, SIWC at 0.2% (w/v) was used in later experiments. Further, it could be concluded that supplementation of colloidal chitin into SIW established induction of chitinase by Btk.

These results are in agreement with most of the previous researches who established the fact that colloidal chitin induced chitinase synthesis by Bt in synthetic medium (Sampson and Gooday, 1998; Rojas-Avelizapa et al., 1999; Barboza-Corona et al., 1999; etc.). Felse and Panda (1999) have also reported that chitinase gene expression in microorganisms has been reported to be controlled by a repressor/inducer system in which chitin or other products of chitin degradation (N-acetylglucosamine) act as inducers. Further, once the available nutrients sources in the medium were exhausted, for survival, Btk would have been forced to induce synthesis of chitinase to hydrolyse chitin in the medium and furnish carbon and nitrogen for growth. Gooday, (1990) and Barboza-Corona et al., (1999) also reported that colloidal chitin could serve as a carbon and nitrogen source but it is not favourable for growth of bacteria in general and Btk in the present case.

2.5.3.2. Fermentor Results

2.5.3.2.1. Fermentation parameters: K_La, OUR and OTR

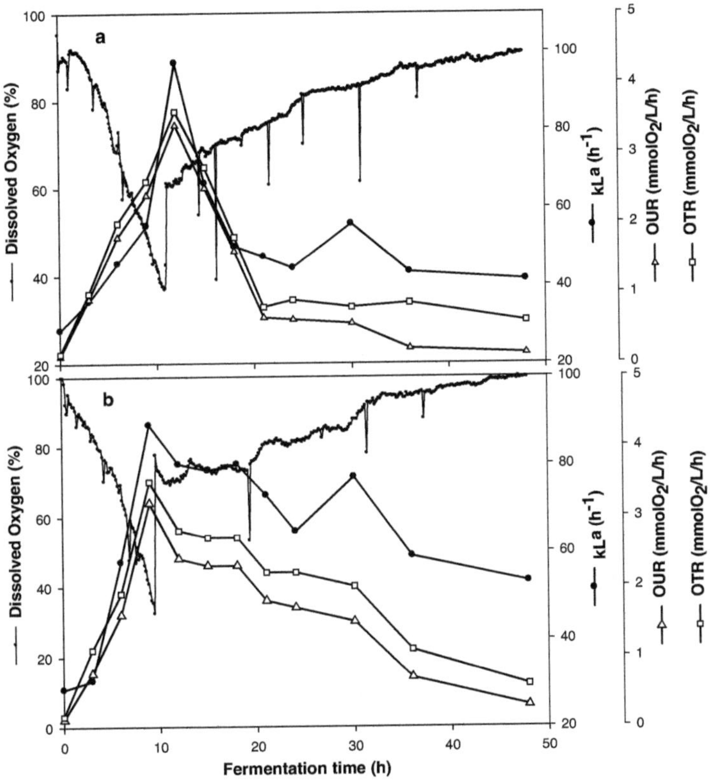

Figure 24. Profiles of Dissolved Oxygen, volumetric oxygen transfer coefficient - k_La, Oxygen Transfer Rate - OTR and Oxygen Uptake Rate - OUR during Btk fermentation of SIW (a) without (control) chitin and (b) fortified with chitin (0.2 % w/v).

To confirm the effect of chitinase as well as the synergistic action of spore, chitinase and delta-endotoxin concentration produced by Btk fermentation on the Tx value, further experiments were conducted in computer-controlled fermentors employing SIW as a sole source of nutrients with and without (control) supplementing colloidal chitin (0.2% w/v) with controlled parameters (DO ≥ 25% of saturation; pH 7, temperature 30°C). The profiles of fermentation parameters (dissolved oxygen -DO, volumetric oxygen transfer coefficient -K_La, oxygen transfer rate -OTR and oxygen uptake rate -OUR) during Btk growth are depicted in Figure 24. The maximum values of fermentation process parameters and Btk growth parameters (specific growth rate, cell and spore counts) obtained in fermentor are presented in Table 18.

During the initial stage of fermentation, DO decreased sharply at 12h and 9h in control and in SIW fortified with chitin, respectively. Then, DO increased and became stable towards the end of fermentation. The decrease in DO was caused by the increase in OUR resulting from the active growth of Btk (exponential growth). The increase in DO might be caused by a decrease in OUR resulting from the stationary phase of Btk as well as the limited nutrients available for growth of Btk. OTR, OUR and K_La increased at the beginning of fermentation and attained their peak at 12h and 9h in control and with chitin, respectively. A decrease in maximum values of K_La, OTR and OUR with chitin (Table 18) was attributed to Btk growth inhibition. Further, K_La, OTR and OUR decreased sharply (below 1 mmolO_2/L/h in approximately 20h) in control run whereas decreased slowly (stayed above 1 mmolO_2/L/h until 36 h of fermentation) with chitin fortification (Fig 24). The total quantity of O_2 used in 48 h of fermentation in SIW (control) (48.9 mmol) was lower than that used in fermentation run fortified with chitin (68.0 mmol). Thus, in spite of lower specific growth rate, low cell and spore counts, higher amount of oxygen was consumed by Btk in presence of chitin. This could be attributed to higer amount of oxygen required for maintenance and cell functioning under inhibitory conditions due to supplementation of chitin.

It should be mentioned that there was appearance of second peak in K_La at 30 h of fermentation process in both bioreactor (Figure 24). This was probably due to the fact that some free spores in the medium germinated and started growing again due to the presence of some residual complex nutrients. The higher value of K_La (second peak) in case of SIWC at 0.2% (w.v) as compared to that in case of SIW (control) indicated that the hydrolysed products of chitin could be used by Btk cells at this period.

2.5.3.2.2. Growth and spore production during the fermentation

The profiles of TC and VS during the Btk fermentation of SIW with and without chitin (control) are presented in Figure 25. The values of TC and SC in case of control were higher than that chitin supplemented SIW. The specific growth rate was substantially decreased whereas the maximum values of OTR and OUR were decreased to a lesser extent and the % sporulation increased with chitin supplementation (Table 18). Lower values of TC, in spite of higher sporulation, also resulted in low spore concentration with chitin. All these facts clearly direct the conclusion towards the fact that added chitin to SIW inhibited growth of Btk as also observed in shake flask experiments. This evidence also supports the contention of Gooday (1990) that chitin is not a favourable source of nutrients for growth of bacteria in general.

Tableau 18. Maximum values of some parameters related to the fermentation process and the growth of Btk in SIW control and SIW + Colloidal Chitin.

Parameters	SIW (control)	SIW + 0.2% Colloidal Chitin
Max. K_La (h^{-1})**	97 h^{-1} (12h)*	89 h^{-1} (9h)*
Max OTR ($mmolO_2/L.h$)**	3.60	3.54
Max. OUR	3.37	3.20

($mmolO_2/L.h$ **		
*Max. specific growth rate (μ max h^{-1})***	0.43	0.34
Max. total cell count (x 10^8 CFU/ml ***	5.60 ± 0.36^{a} (12h)*	4.40 ± 0.26^{b} (15h) *
*Total cell count at 24h (x 10^8 CFU/ml****	3.20 ± 0.21^{a}	2.15 ± 0.13^{b}
Max. spore count at 24h (x 10^8 CFU/ml ***	1.50 ± 0.10^{a}	1.20 ± 0.07^{b}
*Sporulation at 24h (%)****	46.88 ± 0.04^{b}	55.81 ± 0.06^{a}
Total cell count at 48h (x 10^8 CFU/ml ***	2.25 ± 0.14^{a}	1.40 ± 0.09^{b}
Spore count at 48h (x 10^8 CFU/ml ***	1.20 ± 0.07^{a}	1.00 ± 0.07^{b}
*Max. sporulation at 48h (%)****	53.33 ± 0.21^{b}	71.43 ± 0.22^{a}

* Digits in parenthesis represent the time of fermentation at which maximum values of different parameters occurred.

** The presented values are the mean values obtained from two separately experiments conducted for each fermentation condition.

*** The values are the mean of three determinations of two separately experiments conducted for each fermentation condition. The presented valued are the mean ± SD. Different letters (stand near each value) within the same row indicate the significant differences among these values determined by one-factor analysis of variance (Turkey HSD test, $p \leq 0.05$).

2.5.3.2.3. Amylase activity (AA) and alkaline protease activity (PA)

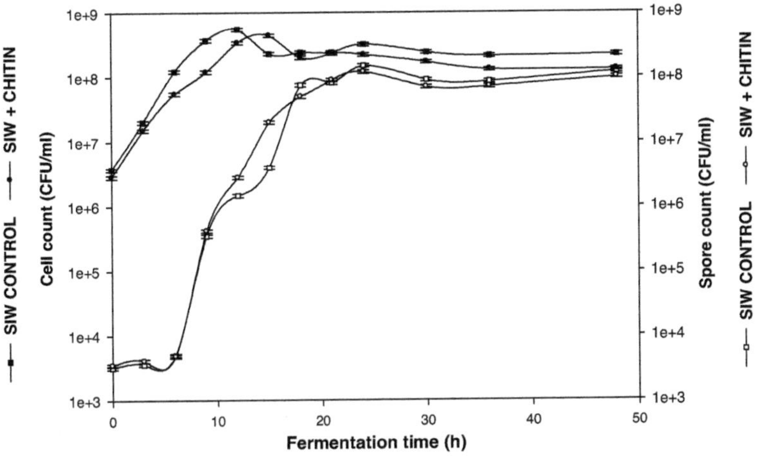

Figure 25. Profiles of total cell count and spore count during Btk fermentation of SIW with and without (control) chitin fortification (0.2 % w/v).

SIW generally contains residual starch and protein (gluten) (Jin et al, 1999) which could serve as substrates inducing the production of amylase and protease, respectively (Zouari and Zaoua, 1999). The profiles of amylase and protease activity during the Btk fermentation of SIW with and without chitin fortification are presented in Figure 26.

There was no substantial difference in protease profiles between control and chitin fortified SIW. In the beginning, PA increased and attained maximum at 15h followed by a decrease until 24h (at this time SC reached near maximum, Table 18). From 6h to 15h of fermentation, presumably, protease was secreted into the medium to hydrolyse complex protein (gluten) into amino acids which were used as nitrogen source for Btk growth (Yezza, 2006; Zouari and Zaoua, 1999). From 15h to 24h, a decrease in protease activity could be possibly due to inactivation of protease. An increase in PA (after 24 h) could be due to the lysis of completely sporulated cells and release of mature spores, insecticidal crystal proteins, cell debris as well as intracellular proteases into the medium (Zouari and Zaoua, 1999). In fact, three autolytic enzymes are associated with sporulation of *Bacillus thuringiensis* and prompt the autolysis of sporulated cells to release the mature spore as well as other components of the cells into the medium (Kingan and Ensign, (1968). They consist of N-acetylmuramidase (pH optimal at 4.0), N-acetylmuramyl-L-alanine amidase (pH optimal at 8.5) and endopeptidase (pH optimal at 8.5). The endopeptidase may be one of the components that cause the increase of PA at the end of fermentation.

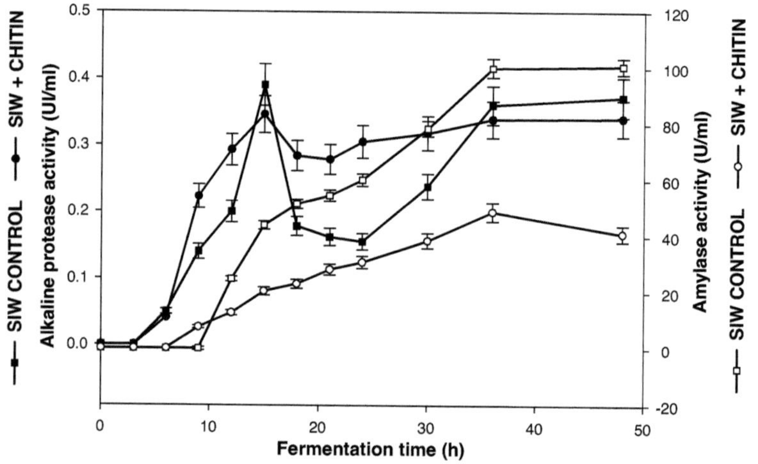

Figure 26. Profiles of alkaline protease activity (IU/ml) and amylase activity (U/ml) during Btk fermentation of SIW with and without (control) fortification of chitin (0.2 % w/v).

There was a substantial difference of AA between control and chitin fortified SIW during the Btk fermentation. AA secretion into the media started from 6h and 9h in chitin fortified SIW and control, respectively and increased until 36h of fermentation. Amylase produced during this period served to hydrolyse the residual starch (available in the media) into glucose that was used as a simple carbon and energy source for Btk growth. The amylase activity produced by Btk in control run was much higher than that of chitin fortified SIW. As discussed above, the supplementation of chitin may become an inhibitor for amylase production while it served as an inducer for chitinase production (discussed later). The hydrolysis of starch to glucose through amylase, in control run, ensured a sustained supply of carbon source (glucose) for growth of Btk until the end of fermentation, whereas due to lower amylase activity in chitin fortified SIW fermentation possibly resulted in a comparative slower supply of glucose and thus leading to higher sporulation than control (71.43% as compared to 53.33%) (Table 18).

2.5.3.2.4. Chitinase activity (CA)

Profiles of chitinase activity during Btk fermentation of SIW with and without chitin fortification (control) are presented in Figure 27. Chitinase activity did not appear in the control whereas with chitin fortification the chitinase activity started increasing at 9h of fermentation and attained maximum value at 24h (15 mU/ml) followed by a decrease until the end of fermentation. Addition of chitinase to SIW induced synthesis of chitinase; however, chitinase activity was not very high. Low chitinase activity has been reported in most of the researches on endogenous chitinase production by Bt. For example, Bt var. *tolworthi* produced chitinase activity of 2 U/ml using shrimp waste medium (Rojas-Avelizapa et al., 1999); Bt var. *kenyae* strains produced chitinase activity from 0.11 to 0.15 U/ml in Castaneda's liquid medium supplemented with chitin (Barboza-Corona et al., 1999); Bt var. *kurstaki* HD-1 (G) produced maximum chitinase activity of 19.30 mU/ml (at 48h of fermentation) using nutrient broth supplemented with 0.3% w/v colloidal chitin (Wiwat et al., 2000); Bt var. *kurstaki* BUPM255 produced maximum chitinase acticity of 28.31 mU/ml at 70 h of incubation in NYSM medium supplemented with 1% w/v colloidal chitin (Driss et al., 2005). The reason for low chitinase activity is due to the fact that chitinase production in *Bacillus thuringiensis* plays a secondary nutrition role and chitin is not an important carbon and nitrogen source (Barboza-Corona et al., 1999).

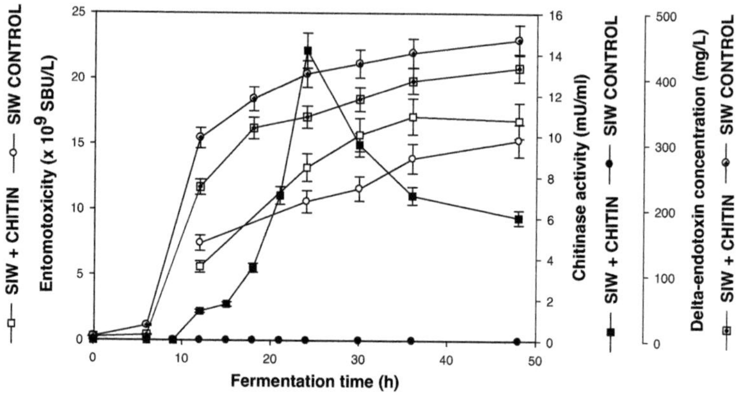

Figure 27. Profiles of chitinase activity, delta-endotoxin concentration and entomotoxicity of fermented broth during Btk fermentation of SIW with and without (control) chitin fortification (0.2 % w/v).

Chitinase activity decreased sharply from 24h to the end of fermentation. The sharp decrease of chitinase activity was also found after 70 h of incubation for Bt var. *kurstaki* BUPM255 (Driss et al., 2005) or after 48h of incubation for Bt var. *kurstaki* HD-1 (G) (Wiwat et al., 2000) when these strains were grown in synthetic medium supplemented with colloidal chitin. The exact reasons for decrease of chitinase activity are not known. However, proteolytic destruction of chitinase has been suggested as a possible reason for this decrease of chitinase activity (Felse and Panda, 1999)

2.5.3.2.5. Delta-endotoxin concentration produced by Btk during the fermentation

Data of delta-endotoxin concentration during Btk fermentation of SIW with and without chitin fortification (control) are presented in Table 19. Delta-endotoxin concentration increased during the course of fermentation concomitantly with spore concentration in control as well as with chitin fortification (Fig 27). A higher delta-endotoxin concentration was observed in control run than with chitin fortification, however the difference in delta-endotoxin concentration at 36 and 48 h in both fermentation processes were not statistically significance. As discussed above, the supplementation of chitin served as the secondary carbon and nitrogen source which was unfavourable for growth of Btk and caused a decrease in spore count and consequently resulted in a slight decrease in delta-endotoxin concentration.

2.5.3.3. Entomotoxicity and the synergistic action of spore, chitinase, dela-endotoxin on bioinsecticidal activity on spruce budworm larvae

2.5.3.3.1. Tx during the fermentation of Btk

Profiles of Tx (against spruce budworm larvae) of fermented broth during Btk fermentation of SIW with and without chitin are presented in Figure 27. It was interesting to note that with chitin fortification, in spite of lower spore count the Tx was higher from 24h to the end of fermentation as compared to the Tx in control. Also, the Tx reached maximum at 36h and 48h with (17.1 x 10^9 SBU/L) and without (15.3 x 10^9 SBU/L) chitin, respectively. The difference in maximum Tx values in both cases was not statistically significant (Figure 27 and Table 19), however, SpTx-spore were significantly different (Table 19). The specific entomotoxicity was also compared based on per mg of delta-endotxin produced (SpTx-toxin) and were found to be significantly different. Since the Tx value with SIWC did not change much between 36 to 48 h, therefore, fermentation with SIWC could be stopped at 36 h.

In case of suspended pellet (complex of spore and delta-endotoxin) in which fermented broth was concentrated 10 times, the Tx values (at 36 and 48 h) of SIWC were significantly higher than Tx values of SIW. There was also a significant difference between SpTx-spores and SpTx-toxin values with and without fortification of chitin (or with and without the presence of chitinase activity) (Table 19). These results clearly demonstrated the role of synergistic action of chitinase with spore and delta-endotoxin in enhancing Tx values.

Tableau 19. Spore count, chitinase acitivty, delta-endotoxin concentration and entomotoxicity at different time of fermentation.

Parameters	Medium	Fermentation time (h)		
		30	36	48
Results fermented broth				
*Spore count (x 10^{10} CFU/L)**	SIW control	9.00 ± 0.51^{b}	8.60 ± 0.52^{b}	12.00 ± 0.70^{a}
	SIW + chitin	7.00 ± 0.42^{c}	7.10 ± 0.40^{c}	9.80 ± 0.55^{b}
*Chitinase activity (U/L)**	SIW control	0^{d}	0^{d}	0^{d}
	SIW + chitin	9.6 ± 0.59^{a}	7.8 ± 0.51^{b}	6.0 ± 0.36^{c}
*Delta-endotoxin concentration (mg/L)**	SIW control	422.62 ± 19.44^{ab}	$439.12 \pm 21,07^{ab}$	459.47 ± 20.70^{a}
	SIW + chitin	392.48 ± 18.84^{b}	401.94 ± 17.28^{b}	407.55 ± 19.15^{ab}
*Entomotoxicity (x 10^{9} SBU/L)**	SIW control	11.6 ± 0.77^{c}	13.9 ± 0.88^{bc}	15.3 ± 0.95^{ab}
	SIW + chitin	15.7 ± 1.02^{ab}	17.1 ± 1.15^{a}	16.8 ± 1.00^{a}
*SpTx-spore (SBU/1000 spore)**	SIW control	128.9 ± 10.3^{c}	161.6 ± 12.9^{bc}	127.5 ± 10.2^{c}
	SIW + chitin	224.3 ± 17.9^{a}	240.8 ± 19.3^{a}	171.4 ± 13.7^{b}

SpTx ($x10^7$ SBU/mg delta-endotoxin) *	SIW control	2.73 ± 0.25^{c}	3.16 ± 0.25^{c}	3.33 ± 0.25^{bc}
	SIW + chitin	4.00 ± 0.30^{ab}	4.26 ± 0.35^{a}	4.10 ± 0.30^{ab}
Results suspended pellet				
Spore count ($x\ 10^{10}$ CFU/L) *	SIW control	85.5 ± 5.6^{b}	81.7 ± 5.3^{bc}	114.0 ± 7.4^{a}
	SIW + chitin	66.5 ± 4.3^{cd}	64.5 ± 4.2^{d}	93.1 ± 6.1^{b}
Chitinase activity (U/L) *	SIW control	0^{d}	0^{d}	0^{d}
	SIW + chitin	7.70 ± 0.50^{a}	6.30 ± 0.42^{b}	5.00 ± 0.31^{c}
Delta-endotoxin concentration (mg/L) **	SIW control	4226.2 ± 194.4^{ab}	4391.2 ± 210.7^{ab}	4594.7 ± 207.0^{a}
	SIW + chitin	3924.8 ± 188.4^{b}	4019.4 ± 172.8^{b}	4075.5 ± 191.5^{ab}
Entomotoxicity ($x\ 10^9$ SBU/L) *	SIW control	17.7 ± 1.4^{c}	19.5 ± 1.6^{bc}	20.5 ± 1.6^{bc}
	SIW + chitin	23.6 ± 1.9^{ab}	26.7 ± 2.1^{a}	25.8 ± 2.1^{a}
SpTx-spore (SBU/1000 spore) *	SIW control	20.7 ± 1.7^{bc}	23.9 ± 1.9^{bc}	17.9 ± 1.4^{c}
	SIW + chitin	38.8 ± 3.1^{a}	39.6 ± 3.2^{a}	25.5 ± 2.0^{b}
SpTx ($x10^6$ SBU/mg delta-endotoxin) *	SIW control	4.20 ± 0.3^{b}	4.40 ± 0.35^{b}	4.45 ± 0.35^{b}
	SIW + chitin	6.56 ± 0.55^{a}	6.64 ± 0.55^{a}	5.84 ± 0.45^{a}

* The values are the mean of three determination of two separately experiments conducted for each fermentation condition. The presented valued are the mean ± SD. Different letters (stand near each value) within the same group (such as spore count, chitinase activity, etc.) indicate the significant differences among these values determined by one-factor analysis of variance (Turkey HSD test, $p \leq 0.05$).

** The values are derived from fermented broth (concentrated 10 times)

To understand the possible mechanisms that caused an increase in Tx with chitin fortification, it was necessary to investigate the synergistic action among some possible factors that influenced the Tx value. Regev et al., 1996 found that there was a quantitative synergistic action of Bt delta-endotoxin CryIC and endochitinase ChiAII against *Spodoptera littoralis*; however, the role of spore was not mentioned. Sampson and Gooday, 1998 found the strong evidence of the importance of endogenous chitinase in the pathogenesis of *Bacillus thuringiensis* by the observation of a decrease in insecticidal activity in the presence of specific chitinase inhibitor allosamidin; however, the delta-endotoxin and spore concentration was not reported. Research of Wiwat et al., 2000 on the toxicity of chitinase – producing Btk (G) toward *Plutella xylostella* was interesting, however, the results of insect toxicity only based on the spore concentration. Other researches focused on screening chitinase producing Bt (Rojas-Avelizapa et al., 1999; Barboza-Corona et al., 1999; Liu et al., 2002); chitinase cloning and characteristic (Wiwat et al. 1996; Tantimavanich et al., 1997; Lertcanawanichakul and Wiwat, 2000; Sirichotpakorn et al, 2001; Tantimavanich et al., 2004; Thamthiankul et al., 2004, etc.), however, the quantitative synergistic action of delta-endotoxin concentration, spore concentration and chitinase activity on the Tx were not also mentioned or evaluated. So, the experiment quantitatively based on these possible synergistic factors affected to overall Tx was investigated in this research.

2.5.3.3.2. Synergistic effects of chitinase and delta-endotoxin and spore on Tx

In fact, the entomotoxicity potential against insect pest larvae depends on many factors such as delta-endotoxin concentration, spore concentration as well as other possible factors such as chitinase activity, protease activity, antibiotic, vegetative insecticidal proteins (Vips) and these factors may act synergistically to impart a Tx value (Aronson, 2002). However, in this research, we only focused on the synergistic action of spore, chitinase activity and delta-endotoxin concentration (Table 19).

In case of fermented broth, the synergistic action of chitinase with delta-endotoxin and spore on Tx was clearly demonstrated in terms of SpTx-spores and SpTx-toxin. In fact, it is known that the main role of Btk based biopesticide in killing insect pest is that of delta-endotoxin and spore (Aronson, 2002). Therefore, the concentration of delta-endotoxin and spore concentration in fermented broth might not be high or concentrated enough to demonstrate clearly the synergistic action with chitinase against spruce budworm.

In case of suspended pellet in which delta-endotoxin and spores were concentrated 10 times, the synergistic action of chitinase with delta-endotoxin and spore against spruce budworm in SIW supplemented with colloidal chitin was demonstrated clearly in comparison to without fortification of chitin (control) or without chitinase (Table 19). Data showed that one of the most important factors that affected the Tx was the delta-endotoxin concentration. The Tx increased concomitantly with delta-endotoxin concentration, especially, when compared the Tx values in suspended pellet and in fermented broth (Fig 27 and Table 19). The second important factor was chitinase activity. In case of chitin supplemented SIW, the delta-endotoxin concentration as well as spore count was lower as compared to those in control (without chitin or without chitinase); but the entomotoxicty was higher (than control) and was possible due to the synergistic action of chitinase. The spores should be the last important factor affecting the overall Tx. The spores is well-known as a synergistic factor towards the overall Tx value (Aronson, 2002), who reported that the insect larvae ingest spores along with the delta-endotoxin and in some cases; spores are synergistic with the toxin. Synergism is probably due to the germination of spores in the midgut with the production of a variety of pathogenic factors by the vegetative cells, including in some subspecies specific vegetative insecticidal proteins, the VIPs (Aronson, 2002). However, the entery of spores into the insect migut could be enhanced by the action of chitinase due to hydrolysis of the peritrophic membrane of insect (Regev et al. 1996). Therefore, it seems that the role of spores is less effective than chitinase with respect to synergistic action on the Tx value against spruce budworm larvae.

2.5.4. Conclusions

Based on the foregoing study of Btk fermentation of satarch industry wastewater with and without (control) fortification of colloidal chitin, the following conclusions could be drawn:

- Total cell count, spore count, alkaline protease and amylase activities were decreased with colloidal chitin supplementation.
- Chitinase activity was induced with colloidal chitin fortification. The maximum value of 15mU/ml of chitinase activity at 0.2% (w/v) concentration of colloidal chitin addition was observed.
- A higher total cell count, spore count and delta-endotoxin concentration was recorded without colloidal chitin supplementation. The entomotoxicity increased with delta-endotoxin concentration with or without supplementation of chitin.
- Maximum entomotoxicity value of suspended pellet (26.7 SBU $x10^9$ SBU/L) reached at 36 h of fermentation with chitin supplementation

whereas it took 48h to attain maximum value (20.5 x 10^9 SBU/L) without colloidal chitin supplementation.

- Specific entomotoxicity with respect to spores and delta-endotoxin were significantly higher with chitin supplementation than without supplementation.
- There was a synergistic action of spore, delta-endotoxin and chitinase activity expressed in terms of entomotoxicity value.

2.5.5. Acknowledgement

The authors are sincerely thankful to the Natural Sciences and Engineering Research Council of Canada (Grant 4984, Canada Research Chair) for financial support. The views and opinions expressed in this article are strictly those of authors and should not be construed as opinions of the U.S. Environmental Protection Agency.

2.5.6. References

1. Aiba, S., Humphrey, A.E., Millis, N.F., 1973. Biochemical Engineering, second ed. Academic Press, New York.
2. Adjalle, K.D., Brar, S.K., Verma, M., Tyagi, R.D., Valero, J.R., Surampalli, R.Y. 2007. Ultrafiltration recovery of entomotoxicity from supernatant of *Bacillus thuringiensis* fermented wastewater and wastewater sludge. Process Biochemistry 42, 1302 - 1311.
3. Annachhatre, A.P., Amatya, P.L., 2000. UASB treatment of tapioca wastewater. Journal of Environmental Engineering. 126. 1149-1152.
4. APHA, AWWA, WPCF, 1998. Standard Methods for Examination of Water and Wastewaters, 20th ed. American Public Health Association, Washington, DC.
5. Aronson, A.I., 2002. Sporulation and δ-endotoxin synthesis by Bacillus thiringiensis. Cellular and Molecular Life Sciences. 59, 417- 425.
6. Arora, N., Ahmad, T., Rajagopal, R., Bhatnagar R.K., 2003. A constitutively expressed 36 kDa exochitinase from *Bacillus thuringiensis*

HD-1. Biochemical and Biophysical Research Communications. 307, 620 - 625.

7. Avignone-Rossa, C., Arcas, J., Mignone C., 1992. *Bacillus thuringiensis*, sporulation and δ-endotoxin production in oxygen limited and nonlimited cultures. World Journal of Microbiology and Biotechnology. 8, 301 - 304.
8. Barboza-Corona, J.E., Contreras, J.C., Velázquez-Robledo, R., Bautista-Justo, M., Gómez-Ramírez, M., Cruz-Camarillo, R., Ibarra, J.E., 1999. Selection of chitinolytic strains of *Bacillus thuringiensis*. Biotechnololy Letters 21, 1125 - 1129.
9. Barboza-Corona, J.E., Nieto-Mazzocco, E., Velázquez-Robledo, R., Salcedo-Hernández, R., Bautista, M., Jiménez, B., Ibarra, J.E., 2003. Cloning, sequencing, and expresion of the chitinase gene chiA74 from *Bacillus thuringiensis*. Applied and Environmental Microbiology. 69, 1023 - 1029.
10. Beegle, C.C., 1990. Bioassay methods for quantification of *Bacillus thuringiensis* δ-endotoxin, Analytical Chemistry of *Bacillus thuringiensis*, in: Hickle, L.A., Fitch, W.L., (Eds.), Analytical Chemistry of *Bacillus thuringiensis*. American Chemical Society, pp. 14–21.
11. Bradford, M.M., 1976. A rapid and sensitive for the quantitation of microgram quantitites of protein utilizing the principle of protein-dye binding. Analytical Biochemistry. 72, 248 - 254.
12. Brar, S.K., 2007. Effets des propriétes rhéologiques sur la fermentation des eaux usées et des boues d'épuration par *Bacillus thuringiensis* var. *kurstaki* et sur le développement de biopesticides en suspensison aqueouses concentrées. Ph.D. thesis, INRS-ETE, Université du Québec, Québec, Canada.
13. Driss, F., Kallassy-Awad, M., Zouari, N., Jaoua, S., 2005. Molecular characterization of a novel chitinase from *Bacillus thuringiensis* subsp. *kurstaki*. Journal of Applied Microbiology. 99, 945 - 953.

14.Dulmage, H.T., Boening, O.P., Rehnborg, C.S., Hansen, G.D., 1971. A proposed standardized bioassay for formulations of *Bacillus thuringiensis* based on the international unit.. Journal of Invertebrate Pathology. 18, 240 - 245.

15.Felse, P.A., Panda T., 1999. Regulation and cloning of microbial chitinase genes. Applied Microbiology and Biotechnology. 51, 141- 151.

16.Gooday, G.W., 1990). Physiology of microbial degradation of chitin and chitosan. Biodegradation. 1, 177 - 190.

17.Guttmann, D.M., Ellar, D.J., 2000. Phenotypic and genotypic comparisons of 23 strains from the Bacillus cereus complex for a selection of known and putative *B. thuringiensis* virulence factors. FEMS Microbiology Letters. 188, 7 - 13.

18.Huang, L.P., Jin, B., Lant P., Zhou, J., 2003. Biotechnological production of lactic acid integrated with potato wastewater treatment by *Rhizopus arrhizus*. Journal of Chemical Technololy and Biotechnology. 78, 899 - 906.

19.Jin, B., van Leeuwen, H.J., Patel, B., Doelle, H.W., Yu, Q., 1999. Production of fungal protein and glucoamylase by *Rhizopus oligosporus* from starch processing wastewater. Process Biochemistry. 34, 59 - 65,

20.Kingan, S.L., Ensign, J.C., 1968. Isolation and characterization of three autolytic enzymes associated with sporulation of *Bacillus thuringiensis* var. *thuringiensis*. Journal of Bacteriology. 96, 629 - 638.

21.Kunitz, M., 1947. Crystalline soybean trypsin inhibitor. Journal of General Physiology. 30, 291 - 310.

22.Lertcanawanichakul, M., Wiwat, C., 2000. Improved shuttle vector for expression of chitinase gene in *Bacillus thuringiensis*. Letters in Applied Microbiology. 31, 123 - 128.

23.Lertcanawanichakul, M., Wiwat, C., Bhumiratana, A., Dean, D.H., 2004. Expression of chitinase-encoding genes in *Bacillus thuringiensis* and toxicity of engineered *B. thuringiensis* subsp. *aizawai* toward *Lymantria dispar* larvae. Current Microbiology. 48, 175 - 181.

24. Lin, Y., Xiong, G., 2004. Molecular cloning and sequence analysis of the chitinase gene from *Bacillus thuringiensis* serovar *alesti*. Biotechnology Letters. 26, 635 - 639.

25. Liu, M., Cai, Q.X., Liu, H.Z., Zhang, B.H., Yan, J.P., Yuan, Z.M., 2002. Chitinolytic activities in *Bacillus thuringiensis* and their synergistic effects on larvicidal activity. Journal of Applied Microbiology. 93, 374 - 379.

26. Miller, G., 1959. Use of dinitrosalicylic acid reagent for determination of reducing sugar. Analytical Chemistry. 31, 426 - 428.

27. Regev, A., Keller, M., Strizhov, N., Sneh, B., Prudovsky, E., Chet, I., Ginzberg, I., Koncz-Kalman, Z., Koncz, C., Schell, J., Zilberstein, A., 1996. Synergistic activity of a *Bacillus thuringiensis* δ-endotoxin and a bacterial endochitinase against *Spodoptera littoralis*. Applied and Environmental Microbiology. 62, 3581 - 3586.

28. Rivera, D., 1998. Growth kinetics of *Bacillus thuringiensis* batch, fed-batch and continous bioreactors cultures. PhD thesis. Faculty of Graduate Studies, The University of Western Ontario, London, Ontario.

29. Roberts, W.K., Selitrennikoff, C.P., 1988. Plant and bacterial chitinases differ in antifungal activity. Journal of General Microbiology. 134, 169 - 176.

30. Rojas-Avelizapa, L.I., Cruz-Camarillo, R., Guerrero, M.I., Rodríguez-Vázquez, R., Ibarra J.E., 1999. Selection and characterization of a proteochitinolytic strain of *Bacillus thuringiensis*, able to grow in shrimp waste media. World Journal of Microbiology and Biotechnology. 15, 299 - 308.

31. Sampson, M.N., Gooday, G.W., 1998. Involvement of chitinases of *Bacillus thuringiensis* during pathogenesis in insects. Microbiology. 144, 2189 - 2194.

32.Sirichotpakorn, N., Rongnoparut, P., Choosang, K., Panbangred, W., 2001. Coexpression of chitinase and the *cry11Aa1* toxin genes in *Bacillus thuringiensis* serovar *israelensis*. Journal of Invertebrate Pathology. 78, 160 - 169.

33.Sklyar, V., Epov, A., Gladchenko, M., Danilovich, D., Kalyuzhnyi, S., 2003. Combined biologic (anaerobic-aerobic) and chemical treatment of starch industry wastewater. Applied Biochemistry and Biotechnology. 109, 253 - 262.

34.Smirnoff, W.A., 1973. Results of tests with *Bacillus thuringiensis* and chitinase on larvae of the spruce budworm. Journal of Invertebrate Pathology. 21, 116 - 118.

35.Smirnoff, W.A., 1974. Three years of aerial field experiments with *Bacillus thuringiensis* plus chitinase formulation against the spruce bud worm. Journal of Invertebrate Pathology. 24, 344 - 348.

36.Tantimavanich, S., Pantowatana, S., Bhumiratana, A., Panbangred, W., 1997. Cloning of a chitinase gene into *Bacillus thuringiensis* subsp. *aizawai* for enhanced insecticidal activity. Journal of General Applied Microbiology. 43, 31 - 37.

37.Thamthiankul, S., Suan-Ngay, S., Tantimavanich, S., Panbangred, W., 2001. Chitinase from *Bacillus thuringiensis* subsp. *pakistani*. Applied Microbiology and Biotechnology. 56, 395 - 401.

38.Thamthiankul, S., Moar, W.J., Miller, M.E., Panbangred, W., 2004. Improving the insecticidal activity of *Bacillus thuringiensis* subsp. *aizawai* against Spodoptera exigua by chromosomal expression of a chitinase gene. Applied Microbiology and Biotechnology. 65, 183 - 192.

39.Tyagi, R.D., Sikati-Foko, V., Barnabé, S., Vidyarthi, A., Valéro, J.R., 2002. Simultaneous production of biopesticide and alkaline proteases by *Bacillus thuringiensis* using wastewater as a raw material. Water Science Technology. 46, 247- 254.

40. Verma, M., Brar, S.K., Tyagi, R.Y., Surampalli, R.Y., Valéro, J.R., 2007. Starch industry wastewater as a substrate for antagonist, *Trichoderma viride* production. Bioresource Technology. 98, 2154 - 2162.
41. Vu, K.D., Tyagi, R.D., Valéro, J.R., Surampalli, R.Y. Impact of different pH control agents on biopesticidal activity of *Bacillus thuringiensis* during the fermentation of starch industry wastewater. Bioprocess Biosyst Eng. DOI 10.1007/s00449-008-0271-z.
42. Wiwat, C., Lertcanawanichakul, M., Siwayapram, P., Pantuwatana, S., Bhumiratana, A., 1996. Expression of chitinase encoding genes from *Aeromonas hydrophila* and *Pseudomonas hydrophila* in *Bacillus thuringiensis* subsp. *israelensis*. Gene. 179, 119 - 126.
43. Wiwat, C., Thaithanum, S., Pantuwatana, S., Bhumiratana, A., 2000. Toxicity of chitinase-producing *Bacillus thuringiensis* ssp. *kurstaki* HD-1 (G) toward *Plutella xylostella*. Journal of Invertebrate Pathology. 76, 270 - 277.
44. Yezza, A., Tyagi, R.D., Valéro, J.R., Surampalli, R.Y., 2006. Bioconversion of industrial wastewater and wastewater sludge into *Bacillus thuringiensis* based biopesticides in pilot fermentor. Bioresource Technology. 97, 1850 - 1857.
45. Zouari, N., Jaoua, S., 1999. Production and characterization of metalloprotease synthesized concomitantly with δ-endotoxin by *Bacillus thuringiensis* subsp. *kurstaki* strain grown on gruel-based media. Enzyme and Microbial Technology. 25, 364 - 371.
46. Zhong, W.F., Jiang, L.H., Yan, W.Z., Cai, P.Z., Zhang, Z.X., Pei, Y., 2003. Cloning and sequencing of chitinase gene from *Bacillus thuringiensis* serovar *israelensis*. Acta Genetica Sinica. 30, 364 - 369.
47. Zhong, W.F., Fang, J.C., Cai, P.Z., Yan, W.Z., Wu, J., Guo, H.F., 2005. Cloning of the *Bacillus thuringiensis* serovar *sotto* chitinase (*Schi*) gene and characterization of its protein. Genetics and Molecular Biology. 28, 821 - 826.

CHAPITRE 3. ACTION DES CHITINASES ET DE LA ZWITTERMICINE A COMME AGENTS DE SYNERGIE DE L'EFFET DES DELTA-ENDOTOXINES ET DES SPORES DE *B. THURINGIENSIS* CONTRE LA TORDEUSE DES BOURGEONS DE L'ÉPINETTE

Partie 3.1. Récupération des chitinases et de Zwittermicine A à partir de surnageant de fermentation de *Bacillus. thuringiensis* (Btk-HD-1) pour produire des biopesticides ayant une très forte activité insecticide

Résumé

Le surnageant de bouillons fermentés par Btk-HD-1 obtenu en employant comme substrat des eaux usées d'industrie d'amidon complétée avec la chitine colloïdale (désigné comme SIWC) et contenant des chitinases et de la Zwittermicine A. a été ultra - filtré sur membrane de 5kDa avec un flux de 480 L/h/m^2 et sous une pression transmembranaire de 75 kPa. Les taux de récupérations de protéines solubles, de chitinases et de spores dans les retentate ont atteint respectivement 79,4%, 50% et 56,7%. La Zwittermicine A a traversé la membrane et a été recueillie dans le filtrat. Le rétentat (contenant les chitinases et autres composants concentrés) et le filtrat ont été mélangés séparément avec des suspensions du culot de SIWC selon divers ratios volumétriques pour obtenir différentes concentrations en delta-endotoxines, en spores et en activité chitinolytique et ces différents mélanges ont été soumis à des bioessais contre la tordeuse des bourgeons de l'épinette les larves (*Choristoreuna fumiferena*). Il a été démontré que les chitinases (et autres composants) ou de la Zwittermicine A ont un effet synergique sur les delta-endotoxine, les spores et l'entomotoxicité (Tx) contre larves de la tordeuse des bourgeons de l'épinette. Le mélange de la suspension du culot et du retentate dans un ratio volumétrique de 1:4 (1Pel4Ret) permet la plus haute récupération de Tx (65,1%) avec le plus gros volume de 5L (1L suspension du culot et 4L retentate). Ce mélange contient des concentrations respectives en spores, en delta-endotoxines et en activité chitinolytique de 3,24 x 10^8 (UFC/ml), 1,09 (mg/ml) et 69,3 (mU/ml). La Tx de ce mélange a été de 22,8 x 10^6 (SBU/ml), soit le ratio le plus efficace en termes de mélange volumétrique. En outre, le mélange de suspension du culot avec du filtrat dans un ratio volumétrique de 1:2 (1Pel2Per) a donné la plus forte action synergique de la Zwittermicine A contenue dans le filtrat avec les delta-endotoxines et les spores sur Tx.

Recovery of chitinases and Zwittermicin A from *Bacillus thuringiensis* fermented broth to produce biopesticide with high biopesticidal activity

Abstract

The *Bacillus thuringiensis* var. *kurstaki* HD-1 (Btk HD-1) fermented broth of starch industry wastewater supplemented with colloidal chitin exhibited the presence of chitinase enzyme and Zwittermicin A. The broth was centrifuged and the supernatant containing chitinase and Zwittermicin A was ultrafiltered employing 5kDa cut-off membrane, operated at feed flux 480 $L/h/m^2$ and transmembrane pressure 75 kPa. Recoveries of soluble protein, chitinase and spore in retentate were 79.4%, 50% and 56.7%, respectively. The Zwittermicin A passed through the membrane and was collected in permeate. The retentate (containing concentrated chitinase) and the permeate were mixed separately with suspended pellet of the fermented broth in different volumetric ratios to obtain different concentrations of delta-endotoxin, spore and chitinase activity and these different mixtures were subjected to bioassay against spruce budworm larvae (*Choristoreuna fumiferena*). Mixing ratio of 4:1 (4 ml retentate: 1ml suspended pellet) demonstrated an effective synergistic action of chitinase (and other possible components) with delta-endotoxin against spruce budworm larvae and thus gave the highest recovery (65 %) of entomotoxicity (Tx) from fermented broth as compared to other mixtures. Similarly, mixing ratio of 2:1 (2 ml permeate: 1 ml suspended pellet) demonstrated an effective synergistic action of Zwitermicin A with delta-endotoxin, spore and chitinase in enhancing the entomotoxicity (Tx) against spruce budworm larvae.

Key words: *Bacillus thuringiensis* var. *kurstaki* HD-1; chitinase; delta-endotoxin; entomotoxicity; synergistic action; ultrafiltration; Zwittermicin A

List of abbreviations

Pel	Suspended pellet
Ret	Retenate
Per	Permeate
Sup	Supernatant
Tx	Entomotoxicity
SBU	Spruce Budworm Unit (bioassay against spruce budworm larvae)
SIW	Starch industry waterwater
SIWC	Starch industry waterwater supplemented with colloidal chitin (0.2%, w/v)
Bt	*Bacillus thuringiensis*
Btk HD-1	*Bacillus thuringiensis* var. *kurstaki* HD-1
SD	Standard deviation
TMP	Transmembrane pressure

3.1.1. Introduction

Chitinase is known to act as a synergistic agent for enhancing the entomotoxicity (Tx) value of biopesticides (Smirnoff, 1973; 1974). It has been suggested that chitinase could increase Tx by perforating the peritrophic membrane barrier in the larval midgut and thus increase the accessibility of the *Bacillus thuringiensis* (Bt) delta-endotoxin molecule to its receptor on epithelial cell membranes (Regev et al., 1996). Perforation of the peritrophic membrane by chitinase was clearly demonstrated in vivo and in vitro examination (Wiwat et al., 2000). These findings confirmed the essential role of chitinase in the hydrolysis of the peritrophic membrane, thus allowing penetration by Bt spores and crystal toxins into the larval hemolymph.

The strain *Bacillus thuringiensis* var. *kurstaki* HD-1 (Btk HD-1) produced chitinase when grown on the semi-synthetic medium supplemented with colloidal chitin (Guttmann et al., 2000). Moreover, it was confirmed that fermentation of Btk using starch industry wastewater supplemented with colloidal chitin (at concentration of 0.2%, w/v) (SIWC) as raw material, Btk HD-1 produced chitinase (Vu et al., 2009). Fermented broth of SIWC demonstrated higher entomotoxicity (Tx) than the Tx of fermented broth of SIW without colloidal chitin supplementation (Vu et al., 2009).

Further, ultrafiltration processes were used as an effective method for recovering many bioactive compounds (enzymes, biopharmaceuticals, etc…) from fermentation broth (Lutz and Raghunath, 2007; Ulbur et al., 2003). Recently, it was demonstrated that the residual delta-endotoxin, spores, cells, and other factors (present in supernatant) of fermented broth of synthetic medium, SIW and wastewater sludge could be recovered by employing ultrafiltration technique, which in turn could be mixed with the centrifugate or pellet (complex of spore and delta-endotoxin) to prepare a high Tx value product (Adjalle et al., 2007). Thus, the chitinase produced by Btk HD-1 fermentation using SIW supplemented with colloidal chitin as raw material, which would normally be discarded in the supernatant, could be recovered during ultrafiltration (in retentate) along with residual delta-endotoxin, spores, cells and other possible synergistic factors (vegetative insecticidal proteins-VIPs and other enzymes). The recovered chitinase through ultrafiltration along with other factors could further enhance the Tx value by mixing the retentate (concentrate obtained from ultrafiltration) with centrifuged pellet in a suitable volumetric ratio.

Moreover, Zwittermicin A is an aminopolyol fungistatic antibiotic that synthesised by many species of *Bacillus* including some strains of Bt, and is produced during sporulation (Stabb et al., 1994; Silo-Suh et al., 1994). It was demonstrated that Zwittermicin A obtained from *Bacillus cereus* UW85 increased the insecticidal activity of commercial Btk HD-1 based biopesticide Foray 76B (a product of Btk HD-1), while demonstrated no insecticidal activity alone when tested against Gypsy moth (*Lymantria dispar*) (Broderick et al., 2000). In fact, Zwittermicin A can be produced by Btk HD-1 in synthetic medium (Stabb et al., 1994; Manker et al., 1994), however, it is not known if Btk HD-1 could produce Zwittermicin A when cultivated in SIW supplemented with colloidal chitin. If Btk could produce Zwittermicin in SIWC fermented broth, due to its low molecular weight of 396 Da (Stabb et al., 1994), Zwittermicin A would be recovered in the permeate fraction (filtrate) of ultrafiltered supernatant employing 5 kDa cut-off membrane. The permeate thus obtained (contain Zwittermicin A) could be mixed with the concentrated delta-endotoxin and spore complex at different ratios to enhance Tx value due to possible synergistic action of Zwittermicin A with delta-endotoxin and spore.

Thus, the objectives of this research were : (1) to confirm the presence of Zwittermicin A in the supernatant of SIW and SIWC fermented broths; (2) fractionation and concentration of chitinase (molecular weight > 30 kDa) (in retentate) and Zwittermicin A (in permeate, molecular weight < 5kDa) from supernatant by ultrafiltration with 5 kDa cut-off membrane; (3) to determine the optimum mixing ratio of retentate (containing chitinase and other components) with pellet (containing spore and delta-endotoxin) to maximise Tx (against spruce budworm larvae) and (4) to determine the optimum mixing ratio of permeate (containing Zwittermicin A) with pellet to maximise the Tx value

3.1.2. Materials and Methods

3.1.2.1. Bacterial strain and biopesticide production medium

Strain Btk HD-1 was taken from stock culture in an agar slant and streaked on tryptic soya agar (TSA) plates, incubated for 48 hour at 30± 1°C and then preserved at 4 °C for future use (Vu et al., 2009).

SIW from ADM-Ogilvie (Candiac, Québec, Canada) was mixed with colloidal chitin (0.2%, w/v) and this mixed solution (SIWC) was used as fermentation medium for Btk based biopesticide production (Vu et al., 2009). Moreover, SIW without supplementation of chitin was also used as raw material for Btk fermentation and served as control to compare the results with SIWC.

3.1.2.2. Fermentation and downstream processing

3.1.2.2.1. Fermentation process

Fermentation was carried out in a stirred tank 15 L fermentor (working volume 10 L, Biogénie, Que., Canada) equipped with accessories and programmable logic control (PLC) system for dissolved oxygen (DO), pH, anti-foam, agitation, aeration rate and temperature. The software (iFix 3.5, Intellution, USA) allowed automatic set-point control and integration of all parameters via PLC. Detail process of fermentation was presented elsewhere (Vu et al., 2009).

3.1.2.2.2. Centrifugation recovery of spore and delta-endotoxin complex from SIW and SIWC fermented broths

At the end of fermentation of SIW and SIWC, the fermented broths were adjusted to pH 4.5 and centrifuged at 9000 g for 30 min to obtain pellets and supernatants (Brar et al., 2006). The pellet of SIW and SIWC fermented broths (after centrifugation) were mixed with their respective supernatant and the volume of each of the suspended pellet thus obtained was one tenth of original volume of the fermented broth, i.e. 10 L of the fermented broth gave 1 L of suspended pellet (designated as Pel) with solids concentration of 71 g/L. The concentration of delta-endotoxin, spore concentrations, chitinase activity and Tx against spruce budworm larvae were determined in the Pel. The total soluble protein, chitinase activity, residual cells, spores and presence of Zwittermicin A concentration were determined in the supernatant of SIWC.

3.1.2.2.3. Ultrafiltration recovery of chitinase and other bioactive compounds from supernant of SIWC

The supernatant obtained after centrifugation of SIWC fermented broth was ultrafiltrated. The equipment used for ultrafiltration was Centramate™ Tangential Flow Systems (Catalogue number OS005C12 - PALL Corporation). The membrane characteristics: 5 kDa cut-off Omega Centramate medium screen; surface area (0.1 m^2) and was made of polyethesulfone; membrane configuration was cassette type (or flat configuration). The process consisted of aseptically feeding 0.5 L of the supernatant called ''feeding solution'' through the membrane to concentrate the bioactive components to a concentrated volume (retentate). The concentration factor was 3.33 (0.5L/0.15L) (Pall Corporation, 2007). The supernatant was brought to room temperature (20-25 °C) and the ultrafiltration process was also conducted at this temperature. The feeding solution was pumped through the membrane at trans membrane pressure (TMP) 75 kPa with flow rate of 48 L/h, which gave a feed flux through the membrane of 480 L/m^2/h. The permeate flux mainly depended on the TMP and the resistance of the membrane. In this study, permeate flux was kept constant (1.50 x 10^{-6} $m^3/m^2/s$) during the ultrafiltration process. This process fractionated supernatant into two fractions: (1) retenate fraction containing concentrated enzymes (chitinase, alkaline protease), spore and other components of higher molecular weight (such as residual delta-endotoxin; Vip3A); (2) permeate fraction containing Zwittermcin A. After ultrafiltration process, permeate and retentate were collected in sterilized flasks. Sampling of the supernatant, retentate and permeate was carried out for analysis of parameters (cells and spores, chitinase activity, total soluble proteins).

3.1.2.3. Analysis parameters

3.1.2.3.1. Chitinase activity (CA)

CA measurement involved colorimetric method (Miller, 1959). CA was determined by measuring the amount of the reducing end group, *N*-acetylglucosamine (NAG), degraded from colloidal chitin, as described by Rojas-Avelizapa et al, (1999) with a modified protocol (Vu et al., 2009). One unit of CA was defined as the amount of enzyme that liberated 1 μmol of NAG per min. The presented values were the mean of three determination of two separate experiments ± SD.

3.1.2.3.2. Identification of specific chitinase activity

Supernatant of SIWC was used to determine the types of chitinase (endo-, exo- or N-acetylglucosaminidase) produced by Btk HD-1 using different specific fluorescent substrates. Different specific fluorescent substrates consisted of 4-Methylumbelliferyl β-D-N,N′,N′-triacetylchitotrioside, 4-Methylumbelliferyl β-D-N,N′,-diacetylchitobioside and 4-Methylumbelliferyl N-acetyl-β-D-glucosaminide were used to determine specific chitinase activity of Endo-chitinase, Exo-chitinase and β-N-acetylhexosaminidase (or N-acetylglucosaminidase), respectively (Sampson and Gooday., 1998). Specific fluorescent substrates for chitinase acticvity were incorporated into agar and poured into Petris. Specific chitinase activities (endo-chitinase, exo-chitinase and β-N-acetylhexosaminidase) were identified by droping an aliquot of enzyme solution (supernatant of fermented broth of SIWC) into a hole which was made on specific fluorescent chitinase substrate-agar, incubating the Petris for 30 min and recording the results under UV Illuminator.

3.1.2.3.3. Estimation of delta-endotoxin concentration produced during the fermentation

Delta-endotoxin concentration was determined based on the solubilisation of insecticidal crystal proteins in alkaline condition (Zouari and Jaoua, 1999). Detailed procedure was presented elsewhere (Vu et al., 2009). The presented values were the mean of three determinations of two separate experiments ± SD.

3.1.2.3.4. Estimation of total cell count (TC) and spore count (SC)

To determine TC and SC, the samples were serially diluted with sterile saline solution (0.85% w/v NaCl). The appropriately diluted samples (0.1mL) were plated on TSA plates and incubated at 30 °C for 24 h to form fully developed colonies. For spore count, the appropriately diluted samples were heated in an oil bath at 80 °C for 10 min and then chilled in ice for 5 min. The TC and SC were performed by counting colonies grown on TSA medium. For cell and spore counts, the average of at least three replicate plates were used for each tested dilution. For enumeration, 30–300 colonies were enumerated per plate. The results were expressed as colony forming units per mL (CFU/ml).

3.1.2.3.5. Determination of the presence of Zwittermicin A in supernatant of SIWC

Bacterial strains and cultural medium. The strain UW85 of *Bacillus cereus* producing Zwittermicin A was used as a positive strain to determine the presence of Zwittermicin A in the supernatant of fermented broth of SIWC. *Bacillus cereus* UW85 was grown with vigorous shaking in half-strength Tryptic soya broth (TSB) for 72h (Stabb et al., 1994; Silo-Suh et al., 1994). Strain *Erwinia herbicola* LS005 was used as bacterial indicator for presence of Zwittermicin A (Stabb et al., 1994). Both strains were kindly supplied by Prof. Haldensman, University of Wincosin, Madison.

The supernatant of SIWC fermented broth was used to determine the presence of Zwittermicin A using bioassay method against *E. herbicola* LS005 (Stabb et al., 1994; Silo-Suh et al., 1994). Supernatant of *Bacillus cereus* UW85 (at 72h) was used as a positive control for producing Zwittermicin A.

Assay for inhibition of growth of E. herbicola. E. herbicola LS005 was inoculated into 5.0 ml of half-strength TSB, and the mixture was incubated overnight at 28°C with vigorous agitation. The culture was diluted in sterile distilled water, and 10^5 cells were spread on a 0.001X Tryptic Soya Agar (TSA) plate. Wells were made in the agar plates by removing a 1.5-cm^2 plug of agar with a flame-sterilized knife. Two hundred micro-litre of bacterial culture (*Bacillus cereus* UW85) or supernatant, retentate and permeate were placed in the wells, and the plates were incubated for 2 days at 21 to 24°C. Plates were scored for the presence or absence of a zone inhibition of growth of *E. herbicola* LS005 around the wells (Stabb et al., 1994; Silo-Suh et al., 1994).

3.1.2.3.6. Optimization of mixing ratios of retentate and permeate with suspended pellet to enhance the Tx

To evaluate the synergistic action of chitinase and other possible components (present in retentate) with complex of spores and delta-endotoxins, the retentate was mixed with the suspended pellet (Pel: Ret, for example 1Pel1Ret implies 1 volume of pellet suspension mixed with 1 volume of retentate) in different volumetric ratios and was subjected to bioassay against spruce budworm larvae to estimate the Tx value. The synergistic effect was evaluated based on control prepared by mixing suspended pellet with saline water (Pel: Sal, volume: volume, 1Pel1Sal implies 1 volume of pellet suspension mixed with 1 volume of saline water) in similar ratios. In order to evaluate the synergistic action of Zwittermicin A (present in the permeate) with complex of spores and delta-endotoxins, the suspended pellet was mixed with the permeate (Pel: Per, volume: volume) in different volumetric ratios and the mixtures were used to determine Tx against spruce budworm larvae. The synergistic effect was evaluated based on control of mixing ratios of Sal: Pel, volume: volume. Further, Tx values of fermented broth SIWC, retentate, permeate and suspended pellet of SIWC (the complex of spores and delta-endotoxins and chitinase) were also determined to compare the results with other mixed samples. The method for estimating Tx based on bioassay technique is presented below.

3.1.2.3.7. Bioassay technique

The Tx of samples was estimated by bioassay against third instar larvae of spruce budworm (Lepidoptera: *Choristoreuna fumiferena*) following the diet incorporation method of Dulmage et al. (1971) and Beegle (1990). The commercial preparation 76B FORAY (Valent Biosciences Corporation Libertyville, IL) was used as a reference standard to analyse the Tx. The detail procedure of bioassay technique was described by Yezza et al. (2006) (Yezza et al., 2006). The Tx was evaluated by comparing the mortality percentages of dilutions of the samples (pellet as well as whole fermented broth) with same dilutions of the standard. In our research, Tx was expressed as relative spruce budworm units (SBU/L). On comparison of Tx value of our samples based on unit of SBU and IU, it was found that SBU reported in our study was 20–25% higher than IU (Yezza et al., 2006). The presented values are the mean of three determination of two independent experiments ± SD.

3.1.3. Results and Discussion

3.1.3.1. Delta-endotoxin, spore concentration, chitinase activity and Tx of Pel

The delta-endotoxin and spore concentration, chitinase activity and Tx value of Pel of SIWC were 5.46 (mg/ml), 1.62 x 10^9 (CFU/ml), 30.6 (mU/ml) and 27.6 x 10^6 (SBU/ml), respectively. Similarly, the obtained pellet of SIW was mixed with supernatant of SIW which did not contain chitinase enzyme (Vu et al., 2008) the Tx value of SIW Pel (no chitinase activity but with almost the same concentration of delta-endotoxin, spore) was 21.0 x 10^6 SBU/ml (data not shown). Thus, SIW Pel was considerably less effective than SIWC Pel. This fact confirmed the role of chitinase activity as a synergistic agent in enhancing the Tx value against spruce budworm larvae as also observed by many previous researchers (Smirnoff, 1973; 1974; Vu et al., 2009). However, it should be mentioned here that the chitinases used by these previous researchers to enhance the Tx value in their biopesticidal formulations came from exogenous sources (animal or microorganisms) i.e. sources other than Bt (Smirnoff, 1973; 1974); whereas, in our research, endogenous chitinase was used (i.e. produced by Btk HD-1 in the fermented broth which was separated in supernatant of SIWC followed by concentrating by ultrafiltraion and mixing with the pellet). This strategy is thus a simple and effective method to produce a Pel of SIWC with higher Tx value.

3.1.3.2. Zwittermicin A and specific chitinase activity in supernatant of SIWC

3.1.3.2.1. Zwittermicin A activity in supernatant of SIWC

The supernatant of SIWC produced comparatively a smaller inhibition zone for E. *herbicola* LS005 growth and supernatant of *Bacillus cereus* UW85 produced a larger zone of inhibition for E. *herbicola* LS005 growth after 24 h of incubation. This fact confirmed the presence of Zwittermicin A in the supernatant of SIWC and thus confirmed that Btk HD-1 could produce Zwittermicin A in SIWC.

3.1.3.2.2. Specific chitinase activity of supernatant of SIWC

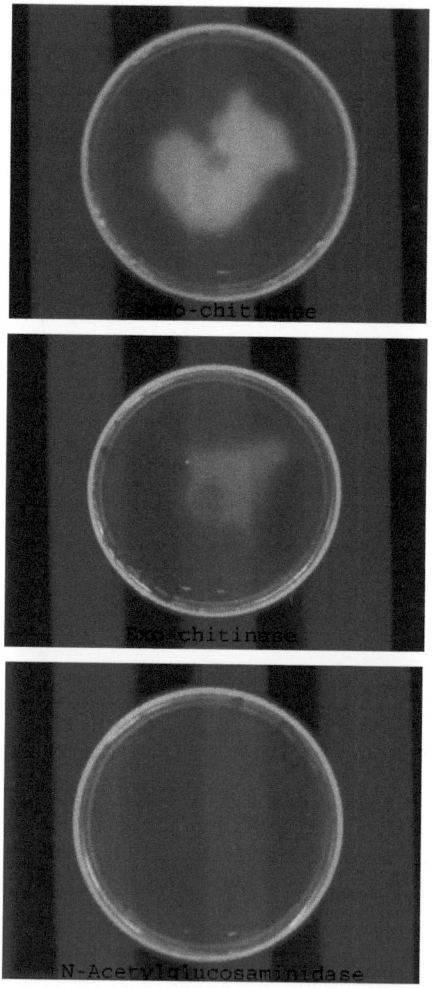

Figure 28. Specific chitinase activity with specific chitinase substrates under UV Illuminator

The appearance of fluorescent light indicates the presence of specific chitinase activity (Figure 28). Results demonstrated that Btk HD-1 produced two types of chitinase while growing in SIW in presence of colloidal chitin: Endo-chitinase and Exo-chitinase in SIWC (Figure 28); which could combine together to hydrolyse the barrier of peritrophic membrane of target insect and therefore could enhance the Tx value However, N-acetylglucosaminidase (or β-N-acetylhexosaminidase) was not found in this supernantant. This meant that Btk HD-1 doesn't produce N-acetylglucosaminidase.

These results are similar to those observed by other workers where it was confirmed that endo-chitinase or exo-chitinase can act synergistically with delta-endotoxin against target insect pests. Regev et al. (1996) cloned and produced Cry1C (encoded by *cry1C* from *Bacillus thuringiensis* K26-21) and an endochitinase ChiA (encoded by *chiAII* from *Serratina marcescens*) in *E.coli* and the authors observed the synergistic activity of a *Bacillus thuringiensis* delta-endotoxin Cry1C and this endochitinase against *Spodoptera littoralis* larvae. Sampson and Gooday (1998) found that 2 strains of *B. thuringiensis* showed high chitinase activities of both endo- and exo-chitinase, particularly for exochitinase.

3.1.3.3. Fractionation and concentration of total soluble protein, chitinase, spores (in retentate) and Zwittermicin A (in permeate) from supernatant of SIWC

The recoveries of total soluble protein, chitinase and spores are presented in Table 20. Results demonstrated that using ultrafiltration system with 5 kDa cut-off membrane, the recovery of protein from supernatant was rather high; however, recoveries of chitinase and spores were only 50 to 57% (Table 20). Chitinase and spores were not found in permeate due to two main reasons: (1) the loss of these components on membrane due to fouling during ultrafiltration which generally occurs on the external membrane surface since most proteins or spores are too large to pass through the pores of the ultrafiltration membranes (Lutz and Raghunath, 2007; Ulbur et al., 2003); (2) the membranes with low molecular weight cut-off ratings (5 kDa in this case) provide higher retention but have corresponding low flux (in this case is 1.50 x 10^{-6} $m^3/m^2/s$); and the low flux results in longer time of ultrafiltration process and probably causes the denaturation of protein components (in this case is chitinase enzymes and other components). This might have finally affected the recovery of chitinase activity in the retentate (Lutz and Raghunath, 2007; Ulbur et al., 2003). To pinpoint the reasons accurately, further research on optimizing transmembrane pressure, verifying protein fouling and membrane resistance during ultrafiltration process need to be conducted.

Moreover, to confirm that Zwittermicin A could be fractionated by ultrafiltration and should be present in permeate only, the obtained fractions of retenatate and permeate from ultrafiltration were used to test the growth inhibition activity against *E. herbicola* LS005. Results demonstrated that only permeate contained Zwittermicin A but retentate did not.

Tableau 20. Recoveries of total soluble protein, chitinase and spore from supernatant of SIWC

Parameters	Supernatant (500 ml)	Retentate (150 ml)	Permeate (350 ml)
Protein concentration (µg/ml)	107.8 ± 5.4	285.5 ± 14.3	n.d *
Protein recovery (%)	100	79.4	n.d
Chitinase activity (mU/ml)	47.6 ± 2.8	79 ± 4.7	n.d
Chitinase recovery (%)	100	50	n.d
Spore concentration (x 10^4 CFU/ml)	8.8 ± 0.7	16.8 ± 1.3	n.d
Spore recovery (%)	100	56.7	n.d
Presence of Zwittermicin A	+	-	+

* n.d. non-detectable

Tableau 21. Synergistic actions of chitinase, Zwittermicin A with delta-endotoxin and spore on Tx

No.	Sample	Spore concentration (x 10^8 CFU/ml)*	Delta-endotoxin (mg/ml)*	Chitinase (mU/ml)*	Tx (x10^6 SBU/ml)*	Total mixing volume (ml) **	Total Tx of mixtures (x 10^9 SBU/L) ***	% recovery Tx from SIWC
1	**1Pel**	**16.2 ± 1.3^{a}**	**5.46 ± 0.27^{a}**	**30.6 ± 1.8^{f}**	**27.6 ± 2.2^{a}**	**1000**	**27.6**	**15.8**
2	1Pel1Sal	8.10 ± 0.6^{d}	2.73 ± 0.14^{d}	15.3 ± 0.9^{i}	17.7 ± 1.4^{ef}	2000	35.4	20.2
3	2Pel1Sal	10.8 ± 0.9^{c}	3.64 ± 0.18^{c}	20.4 ± 1.2^{h}	19.5 ± 1.5^{de}	1500	29.3	16.7
4	**4Pel1Sal**	**13.0 ± 1.1^{b}**	**4.37 ± 0.22^{b}**	**24.8 ± 1.6^{g}**	**20.9 ± 1.7^{cd}**	**1250**	**26.1**	**14.9**
5	1Pel2Sal	5.40 ± 0.4^{e}	1.82 ± 0.09^{e}	10.2 ± 0.6^{j}	16.9 ± 1.4^{ef}	3000	50.7	29.0
6	1Pel4Sal	3.24 ± 0.3^{f}	1.09 ± 0.05^{f}	06.1 ± 0.4^{j}	15.6 ± 1.2^{f}	5000	78	44.6
7	**1Pel1Ret**	**8.10 ± 0.6^{d}**	**2.73 ± 0.14^{d}**	**54.8 ± 3.3^{c}**	**24.4 ± 2.0^{b}**	**2000**	**48.8**	**27.9**
8	**2Pel1Ret**	**10.8 ± 0.9^{c}**	**3.64 ± 0.18^{c}**	**46.7 ± 2.8^{d}**	**28.7 ± 2.3^{a}**	**1500**	**43.1**	**24.6**
9	**4Pel1Ret**	**13.0 ± 1.1^{b}**	**4.37 ± 0.22^{b}**	**40.3 ± 2.4^{e}**	**29.0 ± 2.3^{a}**	**1250**	**36.3**	**20.7**
10	**1Pel2Ret**	**5.40 ± 0.4^{e}**	**1.82 ± 0.09^{e}**	**62.9 ± 3.8^{b}**	**21.7 ± 1.7^{bcd}**	**3000**	**65.1**	**37.2**

11	**1Pel4Ret**	**3.24 ± 0.3 ^{f}**	**1.09 ± 0.05 ^{f}**	**69.3 ± 4.1 ^{a}**	**22.8 ± 1.9 ^{bc}**	**5000**	**114**	**65.1**
12	1Pel1Per	8.10 ± 0.6 ^{d}	2.73 ± 0.14 ^{d}	15.3 ± 0.9 ^{d}	19.6 ± 1.6 ^{de}	2000	39.2	22.4
13	**2Pel1Per**	**10.8 ± 0.9 ^{c}**	**3.64 ± 0.18 ^{c}**	**20.4 ± 1.2 ^{c}**	**21.0 ± 1.7 ^{cd}**	**1500**	**31.5**	**18.0**
14	**4Pel1Per**	**13.0 ± 1.1 ^{b}**	**4.37 ± 0.22 ^{b}**	**24.8 ± 1.6 ^{b}**	**23.2 ± 1.8 ^{bc}**	**1250**	**29**	**16.6**
15	**1Pel2Per**	**5.40 ± 0.4 ^{e}**	**1.82 ± 0.09 ^{e}**	**10.2 ± 0.6 ^{e}**	**22.8 ± 1.8 ^{bc}**	**3000**	**68.4**	**39.1**
16	1Pel4Per	3.24 ± 0.3 ^{f}	1.09 ± 0.05 ^{f}	06.1 ± 0.4 ^{f}	19.0 ± 1.5 ^{de}	5000	95	54.3
17	SIWC broth	1.62 ± 0.1 ^{g}	0.57 ± 0.02 ^{g}	45.0± 2.7 ^{d}	17.5 ± 1.4 ^{ef}	10000	175	100
18	Retentate	0.00168 ± 0.0001	n.d.	79.0 ± 4.7	15.2 ± 1.2 ^{f}	2500	136.8	78.2
19	Permeate	0	0	0	0	6000	0	0

*Different letters (stand near each value) within the same column indicate the Least Significant Differences (LSD) among these values determined by one-factor analysis of variance (Fisher test, $p \leq 0.05$, Statistica 7.0).

** Total mixing volume (ml) based on volume of Pel (1000 ml) + Sal or Per or Ret at different mixing ratios

*** Total Tx of mixtures (x 10^9 SBU/L) = Tx of mixture (x10^6 IU/ml) x total mixing volume (ml)

n.d. not determined

3.1.3.4. Variation of Tx with different volumetric ratios of Tx of suspended pellet (Pel) of SIWC and retentate (Ret) of SIWC

Retentate contained low spore concentration (1.68 x 10^5 CFU/ml) and high chitinase activity (79 mU/ml) (Table 20) and have Tx value of 15.2 x 10^6 SBU/ml (Table 21). This fact demonstrated that retentate also might contain other possible component such as Vips, residual delta-endotoxin and other enzymes (amylases, proteases, and phospholipases). However, so far only chitinases was demonstrated clearly as synergistic agents with delta-endotoxin and spores against the target insect pests (Smirnoff, 1973; 1974; Regev et al., 1996; Wiwat et al., 2000). Other components might contribute to the total Tx of retentate but they are not demonstrated as synergistic agents with delta-endotoxin and spores against the target insect pests: (a) Purified Vip3A from fermented broth of Btk HD-1 demonstrated that Vip3A have low activity against spruce budworm (Milne et al., 2008); (b) the roles of alkaline and neutral proteases were not demonstrated clearly as synergistic agents, it can degrade Cry protoxin but has no effect on the insecticidal activity (Donovan et al., 1997; Tan and Donovan, 2000); (c) phospholipases was known to have activity against some insect pests however its role as synergistic agent was not confirmed (Taguchi et al., 1980). Therefore, in this research, we focused on evaluation the role of chitinase as synergistic agent with delta-endotoxin and spore against spruce budworm.

3.1.3.4.1. Impact of increase of volumetric ratio of Pel in the mixtures (1Pel1Ret; 2Pel1Ret and 4Pel1Ret) on Tx values

The increase in volumetric ratio of Pel in the mixtures (1Pel1Ret; 2Pel1Ret and 4Pel1Ret) lead to the decrease in the concentration of chitinase (and other possible components) and the increase in the concentration of spore and delta-endotoxin (Table 21). The Tx value of these mixtures was increased when concentration of spore, delta-endotoxin increased and chitinase activity decreased (Table 2). Here, the Tx values of the mixtures of 2Pel1Ret (28.7 x 10^6 SBU/ml) and 4Pel1Ret (29.0 x 10^6 SBU/ml) were not statistically different (Table 21); however, the Tx values of these mixtures were statistically higher than the Tx value of the mixture of 1Pel1Ret (24.4x 10^6 SBU/ml). This observation confirmed the principal role of delta-endotoxin and spores on Tx. Moreover, the role of chitinase (and other possible components) on Tx was also apparent from the Tx values of the mixtures of 1Pel1Ret, 2Pel1Ret and 4 Pel1Ret which were significantly higher than that of the Tx values of the mixtures of 1Pel1Sal, 2Pel1Sal and 4 Pel1Sal (control - with same concentration of spore and delta-endotoxin, but relatively lower chitinase activity) (Table 21). Therefore, it could be confirmed that there was synergistic role of chitinase with spores and delta-endotoxin in the mixtures of 1Pel, 2Pel1Ret and 4 Pel1Ret on the Tx value.

3.1.3.4.2. Impact of increase of volumetric ratio of the Retentate (Ret) in the mixtures (1Pel1Ret; 1Pel2Ret and 1Pel4Ret) on Tx values

Increase of volumetric ratio of Ret in the mixtures (1Pel1Ret; 1Pel2Ret and 1Pel4Ret) lead to the increase in concentration of chitinase and the decrease in concentration of spores and delta-endotoxin (Table 21); however the Tx values of these mixtures were not statistically significantly different. Thus, the impact of decrease of spores and delta-endotoxin concentration was probably mitigated by the increase of chitinase activity, consequently, no change of Tx values was observed. These results also demonstrated synergistic role of chitinase (endo- and exo-chitinase) with delta-endotoxin, spores on Tx as also reported earlier through the addition of chitinase to the formulated product (Regev et al., 1996; Sampson and Gooday, 1998). Further, the mixing ratio of 1Pel4Ret was better due to the fact that Tx value of this mixture was equal to those of others and lower volume of the pellet suspension was required.

3.1.3.5. Variation of Tx with different mixing ratios of Pel of SIWC with permeate (Per) of SIWC

It is found that permeate (containing Zwittermicin A) alone doesn't have biopesticidal activity against spruce budworm (Table 21). To verify if

Zwittermicin A present in permeate could have synergistic action with delta-endotoxin, spore and chitinase against spruce budworm, further experiments were conducted in two series. In the first, volumetric ratio of the pellet suspension was increased (1Pel1Per, 2Pel1Per, 4Pel1Ret) whereas in the permeate ratio (1Pel1Per, 1Pel2Per, 1Pel4Per) was increased in the other series. An increase in volumetric ratio of the permeate was expected to increase the concentration of zwittermicin A. The SIWC Pel was used as the main component for mixing at different volumetric ratios with saline water (Sal) – as control. The different mixtures thus obtained were used to determine Tx values against spruce budworm larvae.

3.1.3.5.1. Impact of increase of volumetric ratio of Pel in the mixtures (1Pel1Per; 2Pel1Per and 4 Pel1Per) on Tx value

Results showed that Tx values of these mixtures were increased with increased volumetric ratio of suspended pellet and the increase was mainly due to increased concentration of spore, delta-endotoxin and chitinase activity (Table 21). This fact confirmed the main role of delta-endotoxin, spore and chitinase on Tx (as discussed before).

3.1.3.5.2. Impact of increase of volumetric ratio of Per in the mixtures (1Pel1Per; 1Pel2Per and 1Pel4Per) on Tx values

In these mixtures, concentration of Zwittermicin A was increased with increase in volumetric ratio of permeate (Per) and concentration of spores, delta-endotoxin and chitinase aciticity decreased (Table 21) due to the dilution of suspended pellet (Pel) with permeate (Per). Results revealed that Tx values of the mixture of 1Pel2Per was the highest in three mixtures (1Pel1Per; 1Pel2Per and 1Pel4Per) (statistically significant) (Table 21). It was observed that: (1) concentration of spore, delta-endotoxin and chitinase activity in the mixture of 1Pel1Per was significantly higher than that of 1Pel4Per; (2) concentration of Zwittermicin A in the mixture of 1Pel4Per was higher than that of 1Pel1Per; however, the Tx values of these two mixtures (1Pel4Per and 1 Pel1Per) were not statistically different. This implies that a decrease in spores and delta-endotoxin concentration in 1Pel4Per was compensated by increase in Zwittermicin A concentration (and hence increased the synergistic effect of Zwittermicin A with spores and delta-endotoxin) and therefore the Tx value of this mixture (1Pel4Per) did not decrease even in spite of low concentration of spores and delta-endotoxin. Thus, the mixing ratio of 1Pel2Per was suitable because of higher Tx value than the other two cases.

Further, even if there was synergistic action between Zwittermicin A with delta-endotoxin, spore, and chitinase on Tx against spruce budworm larvae, the mechanism of the synergistic action is not known. In fact, Broderick et al. (2000), suggested that there are several possible mechanisms by which zwittermicin A might enhance Bt activity against Gypsy moth such as: (1) Zwittermicin A may act directly on insect cells once they become accessible through disruption of the midgut epithelium by Bt; (2) Zwittermicin A might also, have a direct effect on midgut functioning, such as disruption of the peritrophic membrane to remove a physical barrier, stimulation of proteases necessary for Bt delta-endotoxin solubilisation and activation, and alteration of midgut epithelium properties to facilitate Bt toxin binding and pore formation; (3) The antimicrobial properties of zwittermicin A may alter the composition of the gut microflora in gypsy moth [12]. Gut microflora are essential for many insect activities, such as normal growth and development, reproduction, digestion, and nutrition and disruption of such relationships could potentially alter the potency of Bt (Broderick et al., 2000).

3.1.3.6. Percentage of recovery of Tx from SIWC

These mixtures prepared above are just the preliminary step in formulation to produce the final biopesticidal products. It was demonstrated that the Tx of final formulated products from these mixtures will be decreased to 22% due to the dilution of mixtures during formulation (decrease in Tx due to dilution caused by addition of various ingredients to formulate the final product) (Brar et al., 2006). It is found that if the Tx values of mixtures are not high and at high volume (for example in cases of using fermented broth of SIWC or Ret of SIWC), when these mixture will be used for formulation, the large volume with low potency after formulation could become a problem in terms of economics due to requirement of large volume of storage and/or transportation. Therefore, the mixtures with high Tx values > 20.0 x 10^6 SBU/ml were chosen as candidates for future formulation.

In case of mixing Pel with Ret, the results demonstrated that the mixture 1Pel4Ret (Tx = 22.8 x 10^6 IU/ml) gave the highest recovery of Tx (65.1% total Tx of SIWC fermented broth) as compared to other cases. Moreover, the total volume of this mixture was 5L which was half of the total volume of SIWC fermented broth (10 L) (Table 21). In cases of mixing Pel with Per, the results also illustrated that the mixture of 1Pel2Per obtained high recovery of Tx (39.1% total Tx of SIWC fermented broth) as compared to other cases (Table 21). Moreover, the obtained total mixing volume of this mixture is 3L which is one-third of the total volume of SIWC fermented broth (10 L) (Table 21). However, the mixture of 1Pel2Per is less effective than the mixture of 1Pel4Ret in terms of total Tx and the (%) recovery of Tx from SIWC fermented broth. Therefore, the mixture 1Pel4Ret could be used efficiently for future formulation to produce final biopesticide product.

3.1.4. Conclusions

Following conclusions can be drawn from the foregoing research:

- Entomotoxicity (27.6 x 10^6 SBU/ml) of suspended pellet obtained after centrifugation of *Bacillus thuringiensis* fermented starch industry wastewater fortified with colloidal chitin (SIWC) was significantly higher than entomotoxicity (21.0 x 10^6 SBU/ml) of the pellet without fortification of chitin.
- Supernatant of fermented broths of SIWC exhibited presence of antibiotic Zwittermicin A and specific chitinase activity (endo-, exo-chitinase activities).

- Using ultrafiltration system with 5 KDa cut-off membrane (flat configuration) at feed flux of 480 $L/m^2/h$ and trans membrane pressure 75 kPa, the recovery of total protein, chitinase activity and spore from supernatant of fermented broth of SIWC was 79%, 50% and 56.7%, respectively.
- There was synergistic action of chitinase or Zwittermicin A with delta-endotoxin and spore on entomotoxicity against spruce budworm larvae.
- Mixing the pellet suspension and the retentate ratio in volumetric ratio of 1:4 (1Pel4Ret) gave the highest recovery of entomotoxicity (65.1%) with highest volume of 5L (1L suspended pellet 4L retentate). This mixture contained spore concentration, delta-endotoxin concentration and chitinase activity of 3.24 x 10^8 (CFU/ml), 1.09 (mg/ml) and 69.3 (mU/ml), respectively with the entomotoxicity of 22.8 x 10^6 (SBU/ml) and was the most effective volumetric mixing ratio in terms of synergistic action of chitinase and delta-endotoxin, spore on entomotoxicity.
- Mixing the pellet suspension with permeate in a volumetric ratio of 1:2 (1PelL pellet and 2L permeate) was the most effective volumetric mixing ratio in terms of synergistic action of Zwittermicin A in permeate with delta –endotoxin and spore on entomotoxicity. This mixture contained spore concentration, delta-endotoxin concentration, chitinase activity and entomotoxicity of 5.40 x 108 (CFU/ml), 1.82 (mg/ml), 10.2 (mU/ml), and 22.8 x 10^6 (SBU/ml) respectively. However, this mixture was less effective in terms of recovery of entomotoxicity from fermented broth compared to the mixture of pellet and retentate.

3.1.5. Acknowledgements

The authors are sincerely thankful to the Natural Sciences and Engineering Research Council of Canada (Grants A4984, and Canada Research Chair) for financial support. Sincere thanks to Prof. Haldensman, University of Wincosin, Madison for providing us *Bacillus cereus* UW85 and *Erwinia herbicola* LS005. The views or opinions expressed in this article are those of the authors.

3.1.6. References

1. Adjalle, K.D., Brar, S.K., Verma, M., Tyagi, R.D., Valero, J.R., Surampalli, R.Y., 2007. Ultrafiltration recovery of entomotoxicity from supernatant of *Bacillus thuringiensis* fermented wastewater and wastewater sludge. Process Biochemistry. 42, 1302 - 1311.
2. Beegle, C.C., 1990. Bioassay methods for quantification of *Bacillus thuringiensis* δ-endotoxin, Analytical Chemistry of *Bacillus thuringiensis*. In: Hickle, L.A., Fitch, W.L. editors. Analytical chemistry of *Bacillus thuringiensis*. USA: American Chemical Society; p. 14–21.
3. Bradford, M.M., 1976. A rapid and sensitive method for the quantitation of microgram quantitites of protein utilizing the principle of protein-dye binding. Analytical Biochemistry. 72, 248-254.
4. Brar, S.K., Verma, M., Tyagi, R.D., Valéro, J.R., Surampalli, R.Y., 2006. Efficient centrifugal recovery of *Bacillus thuringiensis* biopesticides from fermented wastewater and wastewater sludge. Water Research. 40, 1310-1320.
5. Brar, S.K., Verma, M., Tyagi, R.D., Valéro, J.R., Surampalli, R.Y., 2006. Techno-economic analysis of *Bacillus thuringiensis* production process. Final report submitted to INRS-ETE, Research report, No. R-892.
6. Broderick, N.A., Goodman, R.M., Raffa, K.F., Handelsman, J., 2000. Synergy between Zwittermicin A and *Bacillus thuringiensis* subsp. *kurstaki* against Gypsy moth (Lepidoptera: Lymantriidae). Environmental Entomology. 29, 101-107.
7. Donovan, W.P., Tan, Y., Slaney, A.C., 1997. Cloning of the *nprA* gene for neutral protease A of *Bacillus thuringiensis* and effect of in vivo deletion of *nprA* on insecticidal crystal protein. Applied and Environmental Microbiology. 63, 2311-2317.

8. Dulmage, H.T., Boening, O.P., Rehnborg, C.S., Hansen, G.D., 1971. A proposed standardized bioassay for formulations of *Bacillus thuringiensis* based on the international unit. Journal of Invertebrate Pathology. 1971, 240-245.
9. Guttmann, D.M., Ellar, D.J., 2000. Phenotypic and genotypic comparisons of 23 strains from the Bacillus cereus complex for a selection of known and putative *B. thuringiensis* virulence factors. FEMS Microbiololy Letters. 2000, 7-13.
10. Lutz, H., Raghunath, B., 2007. Ultrafiltration process design and implementation. In: Shukla, A.A., Etzel, M.R., Gadam, S., editors. Process scale bioseparations for the biopharmaceutical industry. Florida: Taylor & Francis Group, LLC. p. 297-332.
11. Manker, D.C., Lidster, W.D., Starnes, R.L., MacIntosh, S.C., 1994. Potentiator of *Bacillus* pesticidal activity. Patent Coop Treaty. WO94/09630.
12. Miller, G., 1959 Use of dinitrosalicylic acid reagent for determination of reducing sugar. Analytical Chemistry. 31, 426-428.
13. Milne, R., Liu, Y., Gauthier, D., van Frankenhuyzen, K., 2008. Purification of Vip3Aa from *Bacillus thuringiensis* HD-1 and its contribution to toxicity of HD-1 to spruce budworm (*Choristoneura fumiferana*) and gypsy moth (*Lymantria dispar*) (Lepidoptera). Journal of Invertebrate Pathology. 99, 166-172.
14. Pall Corporation, 2007. Centramate™ & Centramate PE lab tangential flow systems.
15. Regev, A,, Keller, M., Strizhov, N., Sneh, B., Prudovsky, E., Chet, I., Ginzberg, I., Koncz-Kalman, Z., Koncz, C., Schell, J., Zilberstein, A., 1996. Synergistic activity of a *Bacillus thuringiensis* δ-endotoxin and a bacterial endochitinase against *Spodoptera littoralis*. Applied Environmental Microbiology. 62, 3581-3586.

16.Rojas-Avelizapa, L.I., Cruz-Camarillo, R., Guerrero, M.I., Rodríguez-Vázquez, R., Ibarra, J.E., 1999. Selection and characterization of a proteochitinolytic strain of *Bacillus thuringiensis*, able to grow in shrimp waste media. World Journal of Microbiology and Biotechnology. 15, 299 - 308.

17.Sampson, M.N., Gooday, G.W., 1998. Involvement of chitinases of *Bacillus thuringiensis* during pathogenesis in insects. Microbiology. 144, 2189-2194.

18.Silo-Suh, L.A., Lethbridge, B.J., Raffel, S.J., He, H., Clardy, J., Handelsman, J., 1994. Biological activities of two fungistatic antibiotics produced by *Bacillus cereus* UW85. Applied and Environmental Microbiology. 60, 2023-2030.

19.Smirnoff, W.A., 1973. Results of tests with *Bacillus thuringiensis* and chitinase on larvae of the spruce budworm. Journal of Invertebrate Pathology. 21, 116-118.

20.Smirnoff, W.A., 1974. Three years of aerial field experiments with *Bacillus thuringiensis* plus chitinase formulation against the spruce budworm. Journal of Invertebrate Pathology. 24, 344-348.

21.Stabb, E.V., Jaconson, L.M., Handelsman, J., 1994. Zwittermicin A-producing strains of *Bacillus cereus* from diverse soils. Applied and Environmental Microbiology. 60, 4404-4412.

22.Taguchi, R., Asahi, Y., Ikezawa, H., 1980. Purification and properties of phosphatidylinositol-specific phospholipase C of *Bacillus thuringiensis*. Biochimica and Biophysica Acta. 619, 48-57.

23.Tan, Y., Donovan, W.P., 2000. Deletion of *aprA* and *nprA* genes for alkaline protease A and neutral protease A from *Bacillus thuringiensis*: effect on insecticidal crystal proteins. Journal of Biotechnology. 84, 67-72.

24.Ulber, R., Plate, K., Reif, O.W., Melzner, D., 2003. Membranes for protein isolation and purification. In: by Hatti-Kaul R., Mattiasson B.

editors. Isolation and purification of proteins. New York: Marcel Dekker, Inc. p. 191-223.

25. Vu, K.D., Tyagi, R.D., Valéro, J.R., Surampalli, R.Y., 2009 Impact of different pH control agents on biopesticidal activity of *Bacillus thuringiensis* during the fermentation of starch industry wastewater. Bioprocess Biosystems Engineering. 32, 511-519.

26. Vu, K.D., Yan, S., Tyagi, R.D., Valéro, J.R., Surampalli, R.Y., 2009. Induced production of chitinase to enhance entomotoxicity of *Bacillus thuringiensis* employing starch industry wastewater as a substrate. Bioresource Technology. doi:10.1016/j.biortech.2009.03.084

27. Wiwat, C., Thaithanum, S., Pantuwatana, S., Bhumiratana, A., 2000. Toxicity of chitinase-producing *Bacillus thuringiensis* ssp. *kurstaki* HD-1 (G) toward *Plutella xylostella*. Journal of Invertebrate Pathology. 76, 270-277

28. Yezza, A., Tyagi, R.D., Valéro, J.R., Surampalli, R.Y., 2006. Bioconversion of industrial wastewater and wastewater sludge into *Bacillus thuringiensis* based biopesticides in pilot fermentor. Bioresource Technology. 97,1850-1857.

29. Zouari, N., Jaoua, S., 1999. The effect of complex carbon and nitrogen, salt, Tween-80 and acetate on delta-endotoxin production by a *Bacillus thuringiensis* subsp. *kurstaki*. Journal of Industrial Microbiology and Biotechnology. 23, 497-502.

CHAPITRE 4. RELATIONS DE DIVERS PARAMETERS DE BIOPESTICIDES (SPORES, DELTA-ENDOTOXINES ET ENTOMOTOXICITÉ)

Partie 4.1. Relations entre les delta-endotoxines, les spores et l'activité insecticide de biopesticides à base de *Bacillus thuringiensis* (Btk HD-1) produits en utilisant différents milieux de fermentation

Résumé

Les résultats obtenus avec six différents milieux pour la production de *Bacillus thuringiensis* var. *kurstaki* HD-1 (Btk HD-1) ont été analysés. Les concentrations en spores, en delta-endotoxines et l'activité insecticide (entomotoxicité ou Tx) ont été utilisées pour développer les relations possibles entre ces paramètres. La relation entre la concentration de delta-endotoxines et celle des spores dans les différents milieux suit l'équation suivante: concentration de delta-endotoxines = a (concentrations de spores)^b, avec $b > 0$, $R^2 \geq 0.90$; Ceci dépendant de chaque milieu (Les constantes "a et b " sont différentes). La relation entre SpTx-spore (Tx spécifiques par 1000 spores) et la concentration de spores suit strictement l'équation : SpTx-spore = c (concentrations de spores) ^{d}, avec $d < 0$, $R^2 > 0.98$. Ceci conduit à remarquer que (1) des spores produites au début de la période de fermentation peuvent être plus toxiques que celles produites en fin de fermentation, indépendamment des milieux ou (2) Btk HD-1 pourrait produire des agents de synergie (Vip3A, Zwittermicin A, etc), à plus forte concentration au début de la période de fermentation; ces agents contribuant à la valeur de Tx. Ce paramètre et la concentration en Delta-endotoxine suivent une équation linéaire semi-logarithmique.

Mathematical relationships between spore concentrations, delta-endotoxin levels, and entomotoxicity of *Bacillus thuringiensis* preparations produced in different fermentation media

Abstract

The results obtained on six different media used for the production of *Bacillus thuringiensis* var. *kurstaki* HD-1 (Btk HD-1) were analysed in terms of relationship among spore, delta-endotoxin concentrations and biopesticidal activity (entomotoxicity or Tx). Relation between delta-endotoxin and spore concentration and SpTX-spore (specific Tx per 1000 spore) and spore concentration produced in different media strictly followed Power law. The values of constants in power law were characteristics of each medium. Entomotoxicity and delta endotoxin concentration followed the exponential relation whereas a definite relation between entomotoxicity and spore concentration could not be established. It was observed that that the spore and delta-endotoxin produced at the early period of fermentation might be more toxic than those produced during latter period of fermentation irrespective of media used or Btk HD-1 might produce some synergistic agents (Vip3A, Zwittermicin A, etc.) at higher concentration in the early period of fermentation and these agents contributed to enhance Tx value. Tx and delta-endotoxin concentration followed semi-log linear relation. Therefore, delta-endotoxin concentration could be determined rapidly to monitor the progress of the biopesticide production process.

Key words: *Bacillus thuringiensis* var *kurstaki* HD-1, entomotoxicity, semi-synthetic medium, starch industry wastewater, wastewater sludge

4.1.1. INTRODUCTION

Entomotoxicity (Tx) of *Bacillus thuringiensis* (Bt)- based biopesticides against target insects is normally based on a bioassay involving diet incorporation (Dulmage et al. 1970, 1971; Beegle et al. 1990; Brar et al. 2006). The Bt products normally are spore-insecticidal crystal proteins (ICPs) obtained from centrifugation of fermented broth of Bt (Dulmage et al. 1970; Brar et al. 2006). In some cases, to evaluate the insecticidal activity of individual protoxins (delta-endotoxins) such as the 130-140 kDa protoxins (Cry1Aa, Cry1Ab and Cry1Ac) of bipyramidal crystals (component P1) of Bt var *kurstaki* HD-1, a force-feeding method has been used (van Frankenhuyzen et al. 1991). In general, bioassays are laborious, costly and their results depend on insect larvae physiology, diet composition, product formulations and administration (diet incorporation, force-feeding) (Dulmage et al. 1971; Frankenhuyzen et al. 1991). To reduce the need for bioassays, a correlation between alkaline protease activity and Tx was identified (Yezza et al. 2006a).

The main components for killing target insects are delta-endotoxins (Aronson et al. 1986; Frankenhuyzen et al. 1991; Schnepf et al. 1998), but spores and other metabolites of Bt such as vegetative insecticidal proteins (Vips), chitinase, α-exotoxins (phospholipases), immune inhibitor A, Zwittermicin A might also contribute to kill the target insects (Aronson et al. 1986; Schnepf et al. 1998). It is possible that the composition of the media in which Bt is grown influences Tx. Bt var *kurstaki* HD-1 (Btk HD-1) has been grown in starch industry wastewater (SIW) and semi-synthetic medium (Vu et al. 2009a, b, c) and in secondary sludge (Barnabé 2005; Yezza et al. 2005; 2006a, b; Brar et al. 2006). Since in these studies only some of the parameters involved in toxicity to insects were measured or analyzed, the relationships between these parameters (spore concentrations, delta-endotoxin concentrations , entomotoxicity and SpTx-spore) are not exactly known. Therefore, the objectives of this study wereto grow Btk HD-1 in secondary sludge to measure delta-endotoxin and Tx of suspended pellets which were concentrated 10 times by centrifugation and to analyze existing data on production of biopesticides (Btk HD-1) in six different fermentation media (SIW with total solids concentration of 15 g/L, SIW with total solids of 30 g/L, SIW supplemented with 0.2% (w/v) colloidal chitin, SIW supplemented with 1.25 % (w/v) cornstarch and 0.2 % (v/v) Tween 80, secondary sludge, and semi-synthetic medium) to develop relationship between various parameters (spore concentration, delta-endotoxin concentration and Tx value of the fermented broth). These relationships could be used to monitor the progress of biopesticide production process.

4.1.2. MATERIALS AND METHODS

4.12.1. Bacterial strain, inoculum preparation, fermentation medium

Bacillus *thuringiensis* var. *kurstaki* (Btk) HD-1 (ATCC 33679) (Btk HD-1) was cultured on tryptic soya agar (TSA), incubated for 24 h at 30 ± 1°C and preserved at 4 °C for future use (Yezza et al. 2006b).

For the preparation of inoculums for bioreactor cultivations, a loopful of Btk from a TSA plate was inoculated into a 500-ml Erlenmeyer flask containing 100 ml of sterilised tryptic soya broth (TSB) medium, adjusted to pH 7 prior to autoclaving. The flask was incubated on a rotary shaker at 250 revolutions per min (rpm) at 30 °C for 8–12 h. A 2% (v/v) inoculum from this flask was used to inoculate 500-ml Erlenmeyer flasks containing 100 ml of sterilised secondary sludge, semi-synthetic medium or starch industry wastewater (all adjusted to pH 7 prior to autoclaving) and incubated for 8-12 h. Then, the actively growing (exponential growth phase) cells from these flasks were used as an inoculum for bioreactor cultures.

4.1.2.2. Fermentation medium and fermentation procedure

Starch industry wastewater (SIW) was collected from ADM-Ogilvie (Candiac, Québec, Canada) and secondary sludge (SLUDGE) was collected from the wastewater treatment plant- Communauté Urbaine de Québec

(CUQ, Québec, Canada). Semi-synthetic medium (SYN) based on soybean meal was used as control medium. It consisted of (g/L): soybean meal (15.0); glucose (5.0); starch (5.0), K_2HPO_4 (1.0); KH_2PO_4 (1.0); $MgSO_4.7H_2O$ (0.3); $FeSO_4.7H_2O$ (0.02); $ZnSO_4.7H_2O$ (0.02); $CaCO_3$ (1.0) (Yezza et al. 2006b). The list and description of the media discussed in this report is presented in Table 22.

Tableau 22. Media used for cultivation of Bt HD-1

Medium	Description	References
SYN	Semi-synthetic medium	Vu et al., 2009b
SIW30	Starch industry wastewater with total solids (TS) concentration of 30 g/L	Vu et al., 2009b
SIWST	Starch industry wastewater (16.5 g/L TS) + cornstarch (1.25 %, w/v) + Tween 80 (0.2 %, v/v)	Vu et al., unpublished results, 2012a
SLUDGE	Secondary sludge from wastewater treatment plant - Communauté Urbaine du Québec	This study
SIW15	Starch industry wastewater with total solids (TS) concentration of 15 g/L	Vu et al., 2009b
SIWC	Starch industry wastewater (15 g/L TS) + colloidal chitin (0.2%, w/v)	Vu et al., 2009c

Fermentation experiments were conducted in 10 L of media in bioreactors (15 L total capacity) equipped with accessories and automatic control systems for dissolved oxygen, pH, antifoam, impeller speed, aeration rate and temperature. Polypropylene glycol (PPG, Sigma- Canada) (0.1% v/v) solution was added to control foaming during sterilization. The fermenters were sterilized *in situ* at 121 °C for 20 min or for 30 min in case of sludge-containing media. After sterilization, fermenters were cooled to 30 °C and inoculated (2% v/v inoculum) with actively (exponential growth phase) growing cells of the pre-culture. The agitation speed (300–500 rpm) and aeration rate (Litres Par Minute- LPM) were varied in order to keep the dissolved oxygen (DO) values in the range of 30 to 100 % of saturation (Avignone-Rossa et al. 1992). The pH was controlled at 7.0 ± 0.1 using either 4N NaOH or 4 N H_2SO_4 through computer-controlled peristaltic pumps.

4.1.2.3. Preparation of suspended pellet

At the end of fermentation, the pH of the cultures was adjusted to 4.5 and the cultures were centrifuged at 9000 g for 30 min (Brar et al. 2006). The pellets (spore and delta-endotoxin complex) were re-suspended in their respective supernatants to obtain suspension of one-tenth the volume of each original culture. The suspended pellets were used to determine Tx through bioassays.

4.12.4. Analyses

4.1.2.4.1. Estimation of total cell count (TC) and spore count (SC)

The TC and SC were performed by counting colonies grown on TSA medium (Yezza et al. 2006b). For all counts, the average of at least triplicate plates was used for each tested dilution. For enumeration, 30–300 colonies were enumerated per plate. The results were expressed as colony forming units per mL (CFU/ml).

4.1.2.4.2. Estimation of delta-endotoxin concentration produced during the fermentation

Delta-endotoxin concentration was determined based on the solubilization of insecticidal crystal proteins in alkaline condition (Zouari and Jaoua 1999a, b). One ml of samples collected during fermentation was centrifuged at 10 000 g for 10 min. at 4°C. The pellet containing a mixture of spores, insecticidal crystal proteins, cell debris and residual suspended solids was used to estimate the concentration of alkaline-solubilised insecticidal crystal proteins (delta-endotoxin): These pellets were washed three times with 1 ml of 0.14 M NaCl - 0.0 1 % Triton X- 100 solution. The washing helped in eliminating the soluble proteins and proteases which might have adhered on the pellets and could affect the integrity of the crystal protein. The insecticidal crystal proteins in the pellet were solubilized with 0.05N NaOH (pH 12.5) for 3 h at 30°C with stirring. The suspension was centrifuged at 10 000 g for 10 min. at 4°C and the pellet, containing spores, cell debris and residual suspended solids was discarded. The supernatant, containing the alkaline-solubilised insecticidal crystal proteins was used for determination of delta-endotoxin concentration by the Bradford method (Bradford 1976) using bovine serum albumin as standard. The values presented are the average results (± SD) of triplicates of two independent experiments.

4.1.2.4.3. Bioassay

The potential Tx of Btk HD-1 was estimated against third instar larvae of the spruce budworm (*Choristoreuna fumiferena*) following the diet incorporation method (Dulmage et al. 1971; Beegle 1990). The commercial preparation 76B FORAY (Abbott Laboratory, Chicago, IL) was used as a reference standard. Details of the procedure have been described previously (Yezza et al. 2006b). The Tx was evaluated by comparing the mortality percentages of dilutions of the samples with those of the same dilutions of the standard. The Tx was expressed as relative spruce budworm units (SBU/ml). The presented values are the mean of three determination of two independent experiments ± SD.

4.1.2.4.4. Analysis of relationship between parameters

Data obtained from experiments using SYN, SIW15, SIW30, SIWC, SIWST and SLUDGE were used to analyze the relationships between spore concentration and delta-endotoxin concentration, between spore concentration and Tx, and between delta-endotoxin concentration and Tx. All the possible relationships (linear, exponential, or power law) between parameters mentioned above were built using Microsoft Excel. Then, the most accurate relationships between these parameters were chosen for presentation in this report. For all figures, the presented values are the mean of three determinations of two independent experiments ± SD.

4.1.3. RESULTS AND DISCUSSIONS

4.1.3.1. Fermentation of Btk HD-1 using secondary sludge as raw material

Profiles of cell, spore, delta-endotoxin concentration and entomotoxicity (Tx) during the fermentation of Btk HD-1 using sludge as substrate are presented in Fig 29. The maximum concentrations of cell ($5.6x10^8$ CFU/ml) was attained at 15h, whereas those of spores ($4.3x10^8$ CFU/ml) and Tx ($12.4x10^6$ SBU/ml) values were reached at 48 h. Delta-endotoxin concentration reached a maximum value of 582.0 μg/ml. Despite of a similar total solids concentration (30.0 g/L), the maximum delta-endotoxin concentration obtained with sludge (582.0 μg/ml) was significantly lower than that obtained with SIW30 (1112.1 μg/ml) (Fig 30) (Vu et al. 2009b).

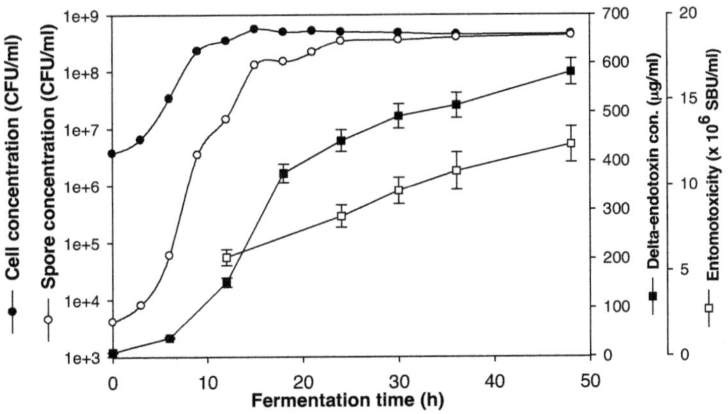

Figure 29. Profiles of cell, spore, delta-endotoxin and Tx during the fermentation of Btk HD-1 using secondary sludge as raw material

* The presented values are the mean of three determinations of two independent experiments ± SD. The medium mentioned in the figure is described in the table 1.

4.1.3.2. Change in Tx, spore and delta-endotoxin concentrations, SpTx-spore and SpTx-toxin during fermentation of Btk HD-1 in different media

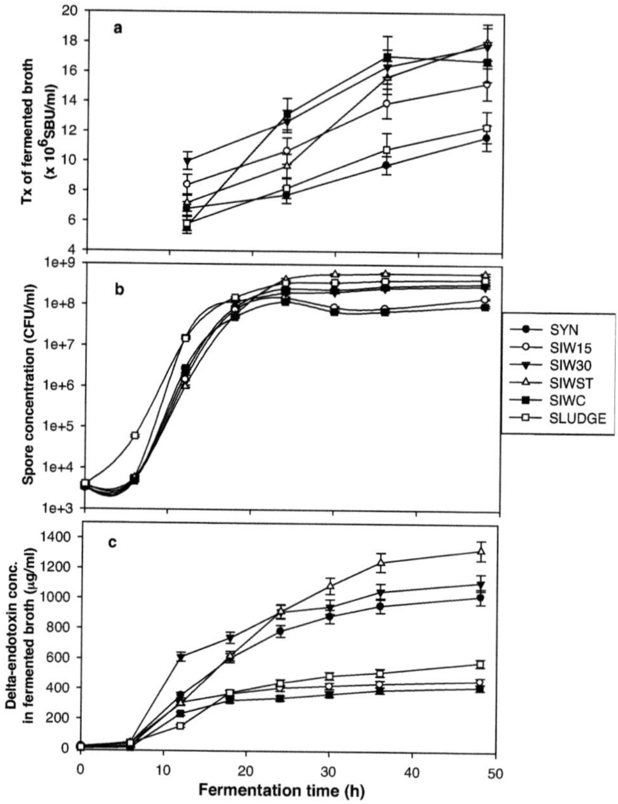

Figure 30. Profile of (a) Tx; (b) spore concentration and (c) delta-endotoxin concentration of fermented broths during the fermentation of different media

* The presented values are the mean of three determinations of two independent experiments ± SD. The media mentioned in the figure are described in the table 1.

The profiles of Tx, spore and delta-endotoxin concentrations are presented in Fig. 30. In general, Tx increased from 12h to the end of fermentation and attained maximum values at the end of fermentation (48h), except in case of SIWC, where Tx was maximum at 36h (Fig. 30a). In most cases, the spore concentration increased from 6 h except in case of SLUDGE, where the spore concentration increased from the beginning of fermentation. Maximum values were obtained at 36 h (SIW30) or at the end of fermentation (SYN, SIWST, SLUDGE) (Fig. 30b). In the case of SIW15 and SIWC, the spore concentration reached maximum at 24 h, decreased slightly from 24h to 36h and increased slightly from 36 h to 48h (Fig. 30b). It is important to note that SIW15 and SIWC had a TS concentration of 15-17 g/L, whereas SYN, SIW30, SIWST and SLUDGE had TS concentration of 28 to 30 g/L.

Delta-endotoxins production by Btk HD-1 increased from 6 h to the end of fermentation. Maximum delta-endotoxin concentrations followed the order: SIWST > SIW30 ≥ SYN > SLUDGE > SIW15 ≥ SIWC ('≥' represents an insignificant difference) (Figure 30c). The reason for the differences is not known, but is likely related to differences in nutrient contents. Delta-endotoxin concentration in SIW15, SIWC and SIW30, increased rapidly until 24 h followed by a slight increase until the end of fermentation. The delta-endotoxin concentrations at 24 and 48 h in SIW15, SIWC and SIW30 were not significantly different; whereas in SYN, SLUDGE and SIWST, the delta-endotoxin concentrations at 24 and 48 h were significantly different.

The profiles of specific entomotoxicity with respect to spores (SpTx-spore, SBU/1000 spores) and specific entomotoxicity with respect to toxin (SpTx-toxin, SBU/μg delta-endotoxin) during fermentation of various media are presented in Fig. 31. The values of SpTx-spore decreased significantly from 12 to 24 h and then stabilised, irrespective of medium (Fig. 31a). This could be due to spores and toxin produced at the early stage of fermentation possessing higher entomotoxic potential against insect larvae or to the production of higher amount of agents contributing towards entomotoxicity (Vip3A, chitinase, vegetative insecticidal proteins-Vips, phospholipases, etc.) at the beginning of cultivation. In fact, Vips are known to be synthesized in the exponential growth phase of Btk HD-1 (Donovan et al. 2001); which generally lasts for 12 h irrespective of the medium used.

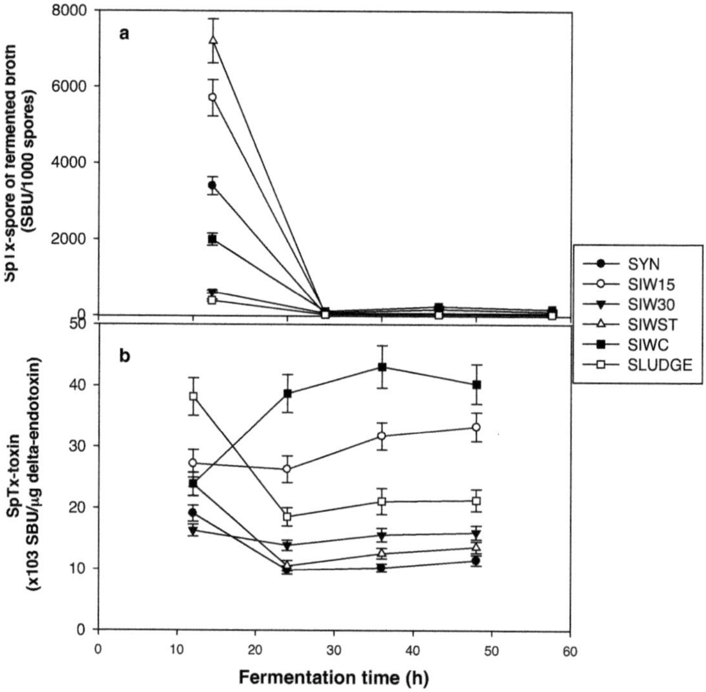

Figure 31. Profiles of (a) SpTx-spore (SBU/1000 spore); (b) SpTx-toxin ($x10^3$ SBU/µg delta-endotoxin) of fermented broths during the fermentation of different media

* The presented values are the mean of three determinations of two independent experiments ± SD. The media mentioned in the figure are described in the table 1.

There were different trends for SpTx-toxin (Tx value/μg delta-endotoxin concentration) values during cultivation in different media. In cases of SYN,

SLUDGE and SIWST, SpTx-toxin values decreased significantly from 12 to 24 h and remained almost constant from 24 to 48 h (Fig. 31b). The highest values of SpTx-toxin occurred at 12 h which could be due to the fact that Btk HD-1 produced a large quantity of Vip3A, phospholipases, etc. (Donovan et al. 2001; Aronson, 1986; Milne et al., 2008). In case of SIW30, the value of SpTx-toxin decreased slightly from 12 to 24 h, and remained mostly unchanged from 24 to 48 h. In case of SIW15, SpTx-toxin did not change from 12 to 24 h; however, it increased significantly from 24 to 48 h. This is in agreement with the observation that the delta-endotoxin concentration in SIW15 did not increase much during the period from 24 to 48 h (Fig. 30c); however, Tx values increased significantly during this period (Fig. 30a). One possible reason could be that Btk HD-1 continued producing soluble agents (Vip3A, Zwittermicin A) from 24 to 48h. The decrease in spore concentration in SIW15 and SIWCbetween 24 to 36 h which might have been due to spore germination (due to the residual nutrients) (Vu et al. 2009a). The germinated spores might begin the growth cycle again and therefore they might produce soluble agents. In case of SIWC, the SpTx-toxin values increased from 12 h (24.5 x 10^3 SBU/ μg delta-endotoxin) to 36h (42.6 x 10^3 SBU/μg delta-endotoxin), perhaps due to synergistic role of chitinases since chitinases activity increased significantly from 9 h (0 mU/ml) to 24 h (15 mU/ml) followed by a significant decrease from 24 to 48 h (6.0 mU/ml) while Tx values increased significantly from 12 h (5.2 x 10^9 SBU/L) to 36 h (17.1 x 10^9 SBU/L) and decreased slightly at 48 h (16.8 x 10^9 SBU/L) (Vu et al. 2009c). SpTx-toxin levels were highest in case of SIWC in spite of the lowest concentrations of spores and delta-endotoxin of all media. It is therefore possible that other soluble factors were produced by Btk HD-1 during the fermentation at different concentrations and these soluble components affected the Tx value and thus caused the changes in the SpTx-toxin value.

4.1.3.3. Relationship between spore and delta-endotoxins and Tx in different media

4.1.3.3.1. Relationship between spore and delta-endotoxin

Relation between delta-endotoxin concentration and spore concentration, irrespective of fermentation media used to produce Btk HD-1, could be expressed by power law (Fig. 32a); DET= $a(SC)^b$, as depicted in Equations 1-6.

Where:

DET: Delta-endotoxin concentration in the medium (µg/ml)

SC: spore concentration in the medium (CFU/ml)

SYN:	$DET = 1.994\ (SC)^{0.318},\ R^2 = 0.96$	(1)
SIW30:	$DET = 1.082\ (SC)^{0.357},\ R^2 = 0.95$	(2)
SIWST :	$DET = 1.832\ (SC)^{0.324},\ R^2 = 0.95$	(3)
SLUDGE :	$DET = 0.674\ (SC)^{0.335},\ R^2 = 0.98$	(4)
SIW15:	$DET = 0.758\ (SC)^{0.349},\ R^2 = 0.90$	(5)
SIWC:	$DET = 0.250\ (SC)^{0.406},\ R^2 = 0.96$	(6)

The greater the value of "b", the higher the delta-endotoxin concentration at a given spore concentration. The factor "a" is a direct measure of delta-endotoxin (proportionality constant between spore concentration and delta-endotoxin concentration) concentration at a given spore concentration. The values of constant 'b' in different media followed the order: SIWC > SIW30 > SIW15 > SLUDGE > SIWST > SYN and therefore, when taking the 'power function' of the same concentration of spore with different exponent constant 'b' of these media, the values will also follow the above order; however, the value for 'a'were small ($a < 1$) for SIWC, SIW15 and SLUDGE. The values of constants 'a and b' for SYN, SIWST, SLUDGE, and SIW15 were not much different as spore and delta-endotoxin production was high. The smallest values of 'a' occurred in case of SIWC, which had the lowest concentrations of delta-endotoxin and spore presumably due to inhibition by added colloidal chitin (Vu et al. 2009c). In case of SIW15, delta-endotoxin and spore concentrations were low causing low values of 'a'. In case of SLUDGE, the spore concentration produced was higher than in SIW15; however, delta-endotoxin produced in SLUDGE was not significantly higher than that in SIW15 (Fig. 30), causing a lower value of 'a' for SLUDGE than for SIW15. The lower spore or delta-endotoxin concentrations in SLUDGE or SIW15 could be due to the lower available nutrients compared to SYN, SIW30 or SIWST, in agreement with the hypothesis that SLUDGE (without treatment) or SIW15 (without any modification) may not have enough nutrients for growth, sporulation and delta-endotoxin synthesis by Btk HD-1(Barnabé 2005; Yezza et al. 2005; Vu et al. 2009a, b; Vu et al., 2012a).

In fact, when the concentration of crystals produced by Bt is known, the approximate concentration of protoxins (delta-endotoxins) can be estimated based on the assumption that one crystal (insecticidal crystal proteins) consist of $1x\ 10^6$ - $2x10^6$ delta-endotoxin molecules (Agaisse and Lereclus 1995) and that the molecular weight of one delta-endotoxin is about 130-140 kDa (Aronson et al. 1986). However, it is difficult to estimate the concentration of crystals based on spore concentration because Btk HD-1 can produce two inclucions in one sporulated cell during sporulation, a bipyramidal inclusion (consisting of 130-140 kDa protoxins, delta-endotoxins, active against lepidopteran larvae) and an ovoid inclusion (consisting of 65 kDa protoxins, active against both lepidopteran and dipteran larvae) (Aronson et al. 1986) and because some cells lacked ovoid inclusions and more than one bipyramidal parasporal crystal can develop within a cell, but only one ovoid inclusion per sporangium has been observed (Bechtel and Bulla 1976). Also, the sporulation process was not synchronous because spores were observed from the beginning of fermentation process (at 0h). Thus, there might have been crystals in the medium during the fermentation and hence the number of crystals produced based on spore count can not be calculated.

4.1.3.3.2. Relationship between Tx and spore concentration

Tx as a function of spore concentration in different media is presented in Fig. 32b. There was no general relationship between Tx and spore concentration. In few cases the relationship could be expressed by the exponential function ($Tx = ge^{f(SC)}$), as shown in Equations 7-12.

SYN:	$Tx = 6.17e^{1.61E\text{-}9(SC)}$, $R^2 = 0.745$	(7)
SIW30:	$Tx = 9.07e^{2.18E\text{-}9(SC)}$, $R^2 = 0.935$	(8)
SIWST:	$Tx = 5.79e^{1.72E\text{-}9(SC)}$; $R^2 = 0.863$	(9)
SLUDGE:	$Tx = 5.07e^{1.88E\text{-}9(SC)}$, $R^2 = 0.884$	(10)
SIW15:	$Tx = 9.58e^{2.31E\text{-}9(SC)}$; $R^2 = 0.389$	(11)
SIWC:	$Tx = 8.66^{5.40E\text{-}9(SC)}$; $R^2 = 0.482$	(12)

For SIW30, the relationship followed the exponential function with R^2 = 0.935. The exponential fit of the data in cases of SLUDGE and SIWST was acceptable. In previous research, the relation between Tx and spore concentration during wastewater sludge fermentation at different suspended solids concentration (15, 20 and 25 g/L) also followed the exponential function ($R^2 = 0.9116$) (Yezza et al., 2006a). In the cases of SYN, SIW15 and SIWC, the exponential fit for relationship between Tx and spore concentration was poor ($R^2 \leq 0.747$). These observations also confirm the observation by Yezza et al. (2006a) that the relation between Tx and spore concentration changes depending on the culture medium, possibly because Tx value is not only dependent on spore concentrations but also the delta-endotoxin concentrations and other soluble factors (Vip, Zwittermicin A, etc.).

To further confirm that the relationship between Tx and spore concentration was also dependent on other factors, the data of spore concentration and Tx value were re-analysed in terms of specific entomotoxicity with respect to spores (spTx-spore, SBU/1000 spore) and spore concentration (obtained at different time of fermentation for different media) (Fig. 32c). The relation followed power law in all cases (Equation 13-18).

Where:

SpTx-spore: specific entomotoxicitywith respect to spores

SYN:	SpTx-spore = $2x10^9 (SC)^{-0.9272}$, $R^2 = 0.995$	(13)
SIW30:	SpTx-spore = $6x10^8 (SC)^{-0.8305}$, $R^2 = 0.990$	(14)
SIWST:	SpTx-spore = $2x10^9 (SC)^{-0.8897}$, $R^2= 0.992$	(15)
SLUDGE:	SpTx-spore = $3x10^8 (SC)^{-0.8184}$, $R^2 = 0.987$	(16)
SIW15:	SpTx-spore = $2x10^9 (SC)^{-0.8999}$, $R^2 = 0.994$	(17)
SIWC:	SpTx-spore = $9x10^7 (SC)^{-0.7202}$, $R^2= 0.982$	(18)

SpTx-Spore decreased rapidly with spore concentration in the beginning of fermentation (or at lower spore concentrations), remained constant at higher spore concentrations (Fig 32c). It is possible that 1) spore and crystals produced in the beginning were more toxic than those produced at a later stage of fermentation; 2) some of the spores germinated due to a high concentration of nutrients in the beginning and the actual spore count decreased; and 3) the concentration of VIPs, Zwittermycin A and other soluble components produced in large quantity in the beginning of cultivation contributed towards a high SpTx-spore value.

Zwittermicin A is known to be synthesized during the sporulation of *Bacillus cereus* and its synthesis is dependent on components of the medium (Milner et al. 1995). Zwittermicin A is also produced by Btk HD-1 in SIW medium fortified with chitin or SIWC (Vu et al. 2012b).

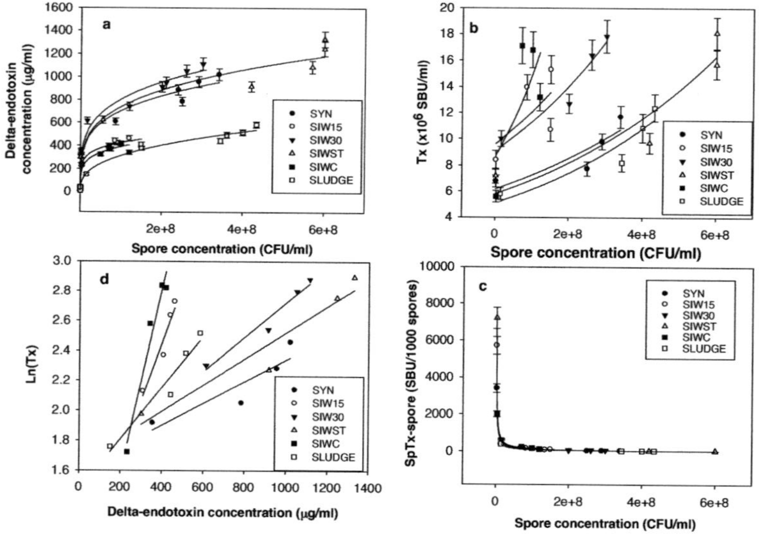

Figure 32. (a) Relations between (a) spore and delta-endotoxin concentration; (b) Tx and spore concentration; (c) specific Tx –spore (SBU/1000 spore) and spore concentration; (d) Ln(Tx) and Delta-endotoxin concentration of different fermented broths

* The presented values are the mean of three determinations of two independent experiments ± SD. The media mentioned in the figure are described in the table 1.

4.3.3.3. Relationship between delta-endotoxin and Tx

Relationships between Tx and delta-endotoxin concentrations (μg/ml) (observed at different time during Btk fermentation of different media) were found to be exponential and are presented on semi log plots in Fig. 32d and expressed as Equations 19-24.

SYN:	$\text{Ln (Tx)} = 7.3\text{x}10^{-4}\ (\text{DET}) + 1.61,\ R^2 = 0.817$	(19)
SIW30:	$\text{Ln (Tx)} = 1.1\text{x}10^{-3}\ (\text{DET}) + 1.56,\ R^2 = 0.959$	(20)
SIWST:	$\text{Ln (Tx)} = 8.8\text{x}10^{-4}\ (\text{DET}) + 1.64,\ R^2 = 0.920$	(21)
SLUDGE:	$\text{Ln (Tx)} = 1.7\text{x}10^{-3}\ (\text{DET}) + 1.46;\ R^2 = 0.942$	(22)
SIW15:	$\text{Ln (Tx)} = 3.9\text{x}10^{-3}\ (\text{DET}) + 0.88,\ R^2 = 0.929$	(23)
SIWC:	$\text{Ln (Tx)} = 6.3\text{x}10^{-3}\ (\text{DET}) + 0.30,\ R^2 = 0.960$	(24)

The semi-log relationship was highly correlated ($R^2 > 0.92$), except for SYN where $R^2 > 0.81$.

The linear relations between Ln (Tx) and delta-endotoxin concentration could be used to follow the follow the progress of Bt-based biopesticide production process by measuring delta-endotoxin concentrations, a process that can be done rapidly. However, these linear equations should be used only for monitoring the fermentation process or for prediction the Tx that can be obtained at the end of fermentation process. The best way for having the exact Tx value is by bioassay due to the presence of different crystal forms with different insecticidal spectra and in different ratios (Bechtel and Bulla 1976, Aronson et al. 1986) and variability in components contributing to toxicity.

4.1.3.4. Comparison of Tx values, spore and delta-endotoxin concentrations, SpTx-spore and SpTx-toxin of suspended pellets from different media

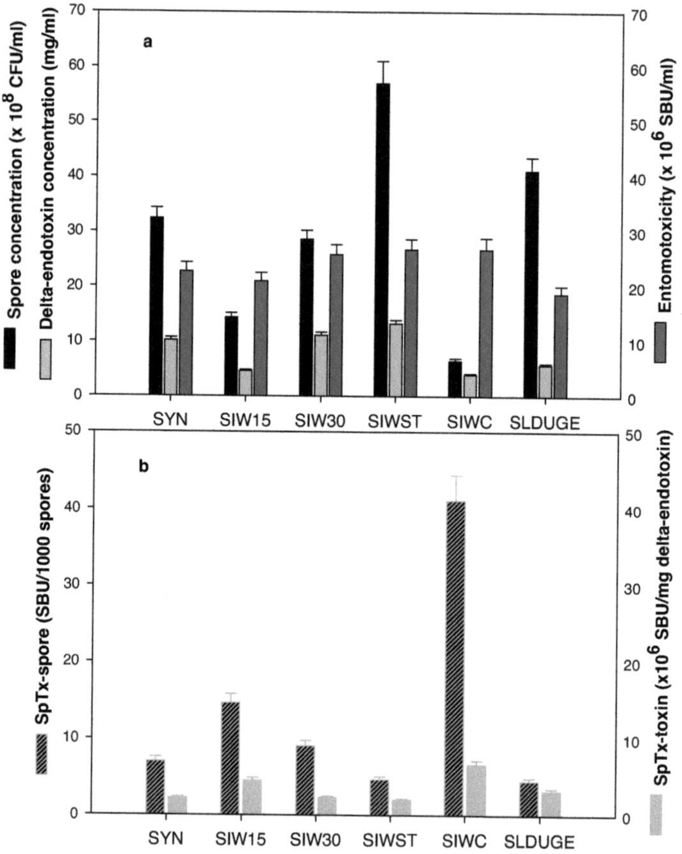

Figure 33. Profiles of (a) Spore, delta-endotoxin concentrations and entomotoxicity; (b) SpTx-spore and SpTx-toxin of suspended pellets obtained from different fermentation media

* The presented values are the mean of three determinations of two independent experiments ± SD. The medium mentioned in the figure is described in the table 1

Spore and delta-endotoxin concentrations, Tx values, SpTx-spore (SBU/1000 spores) and SpTx-toxin (x10^6 SBU/mg delta-endotoxin) of suspended pellets from the six fermented media are presented in Fig. 33. Spore and delta-endotoxin concentrations in case of SIWST were significantly higher than those of SIWC; however, their Tx values were not significantly different. This was due to the presence of chitinases in suspended pellet of SIWC, which acted synergistically to augment the Tx value (Vu et al., 2009d). Further, spore and delta-endotoxin concentrations of SIWST were significantly higher than those of SIW30, but their Tx values were not significantly different. Since supernatant was used to re-suspend the respective pellets it is possible that the soluble factors (Vips, Zwittermicin A, etc.) present in supernatants might have contributed differently to Tx values of these suspended pellets. The spore concentration of SYN was higher than that of SIW30 but, on the contrary, delta-endotoxin concentration of SIW30 was higher than that of SYN and that finally caused Tx to be higher in SIW30 than in SYN (Fig. 33). Finally, Tx values of SYN, SIW15 and SLDUGE were not significantly different in spite of differences in spore or delta-endotoxin concentrations.

The values of SpTx-spore (SBU/1000 spores) and SpTx-toxin (x 10^6 SBU/mg delta-endotoxin) were highest in case of SIWC due to high Tx values at the lowest concentrations of spores and delta-endotoxin and was due to the presence of chitinases. In cases of SYN, SIW30 and SIWST, SpTx-spore values were different but SpTx-toxin values were not significantly different which demonstrated the principal role of delta-endotoxin in these cases in killing the insects. The values of SpTx-spore and SpTx-toxin of suspended pellets of SIW15 and SLUDGE were higher than those of SYN, SIW30, and SIWST, possibly due to the presence of higher concentration of soluble factors in suspended pellets of SIW15 and SLUDGE than in SYN, SIW30 and SIWST.

4.1.4. CONCLUSIONS

Analysis of data obtained from the cultivation of Btk HD-1 in different media was conducted to find relationships that may exist between three most important parameters (spore concentration, delta-endotoxin concentration and entomotoxicity) of any Bt-based preparations. The relationship between delta-endotoxin and spore concentration produced by Btk HD-1 followed the Power law, irrespective of growth medium. Thus, it is possible to predict the delta-endotoxin concentrations based on spore concentrations in cultivation media. The relationships between entomotoxicity and delta-endotoxin concentrations of Btk HD-1 produced in different media were found to be linear on semi log scale. These relationships indicate that delta-endotoxin concentration can be used as an indicator for monitoring the fermentation process and for predicting the possible entomotoxicity of cultured media of Btk HD-1. However, bioassay is always required to be conducted to have exact values of entomotoxicity of biopesticide preparation against target insects. Future research on analysis of soluble metabolites (vegetable insecticidal proteins, Zwitermicin A, chitinase, etc.) produced by Btk HD-1 during fermentation process in different media should be conducted to further explore the roles of these components on overall entomotoxicity.

4.1.5. ACKNOWLEDGEMENTS

The authors are sincerely thankful to the Natural Sciences and Engineering Research Council of Canada (Grant A4984, Canada Research Chair) for financial support. The authors sincerely thank **ADM-Ogilvie** (Candiac, Québec, Canada) for providing starch industry wastewater and thank **Insect Production Unit, Great Lakes Forest Research Centre** (Sault Ste. Marie, Ontario) for providing spruce budworm larvae during conducting this project The views and opinions expressed in this article are strictly those of authors.

4.1.6. References

1. Agaisse H, Lereclus D (1995) How Does *Bacillus thuringiensis* Produce So Much Insecticidal Crystal Protein? J Bacteriol 177: 6027–6032.
2. Aronson AI, Beckman W, Dunn P (1986) *Bacillus thuringiensis* and related insect pathogens. Microbiol Rev 50: 1-24.
3. Avignone-Rossa C, Arcas J, Mignone C (1992) *Bacillus thuringiensis*, sporulation and δ-endotoxin production in oxygen limited and nonlimited cultures. World J Microbiol Biotechnol 8: 301–304.
4. Barnabé S (2005) Hydrolyze et oxydation partielle des boues d'épuration comme substrat pour produire *Bacillus thuringiensis* HD-1. Ph.D thesis, INRS - Université du Québec, Québec, Canada.
5. Bechtel D, Bulla LAJr (1976) Electron microscope study of sporulation and parasporal crystal formation in *Bacillus thuringiensis*. J Bacteriol 127: 1472-1481.
6. Beegle CC (1990) Bioassay methods for quantification of *Bacillus thuringiensis* δ-endotoxin In:. Hickle LA, Fitch WL (eds) Analytical Chemistry of *Bacillus thuringiensis*. American Chemical Society, New York, pp. 14–21.
7. Bradford MM (1976) A rapid and sensitive method for the quantitation of microgram quantitites of protein utilizing the principle of protein-dye binding. Anal Biochem 72: 248-254.
8. Brar SK, Verma M, Tyagi RD, Valéro JR, Surampalli RY (2006) Efficient centrifugal recovery of *Bacillus thuringiensis* biopesticides from fermented wastewater and wastewater sludge. Water Res 40:1310-1320.
9. Donovan WP, Donovan JC, Engleman JT (2001) Gene knockout demonstrates hat *vip3A* contributes to the pathogenesis of *Bacillus thuringiensis* toward *Agrotis ipsilon* and *Spodoptera frugiperda*. J Invertebr Pathol 78: 45–51.

10. Dulmage HT, Correa JA, Martinez AJ (1970) Coprecipitation with lactose as a means of recovering the spore-crystal complex of *Bacillus thuringiensis*. J Invertebr Pathol 15: 15-20.
11. Dulmage HT, Boening OP, Rehnborg CS, Hansen GD (1971) A proposed standardized bioassay for formulations of *Bacillus thuringiensis* based on the international unit. J Invertebr Pathol 18: 240-245.
12. Milne R, Liu Y, Gauthier D, van Frankenhuyzen K (2008) Purification of Vip3Aa from *Bacillus thuringiensis* HD-1 and its contribution to toxicity of HD-1 to spruce budworm (*Choristoneura fumiferana*) and gypsy moth (*Lymantria dispar*) (Lepidoptera). J Invertebr Pathol 99: 166-172.
13. Milner JL, Raffel SJ, Lethbridge BJ, Handelsman J (1995) Culture conditions that influence accumulation of zwittermicin A by *Bacillus cereus* UW85. Appl Microbiol Biotechnol 43: 685-691.
14. Schnepf E, Crickmore N, van Rie J, Lereclus D, Baum J, Feitelson J, Zeigler DR, Dean DH (1998) *Bacillus thuringiensis* and its pesticidal crystal proteins. Microbiol Mol Biol Rev 62: 775-806.
15. Stabb EV, Jaconson LM, Handelsman J (1994) Zwittermicin A-producing strains of *Bacillus cereus* from diverse soils. Appl Environ Microbiol 60: 4404-4412.
16. van Frankenhuyzen K, Gringorten JL, Milne RE, Gauthier D, Pusztai M, Brousseau R, Masson L (1991) Specificity of activated CryIA proteins from *Bacillus thuringiensis* subsp. *kurstaki* HD-1 for defoliating forest Lepidoptera. Appl Environ Microbiol 57: 1650-1655.
17. Vu KD, Tyagi RD, Valéro JR, Surampalli RY (2009a) Impact of different pH control agents on biopesticidal activity of *Bacillus thuringiensis* during the fermentation of starch industry wastewater. Bioprocess Biosyst Eng 32: 511-519.

18. Vu KD, Tyagi RD, Brar SK, Valéro JR, Surampalli RY (2009b) Starch industry wastewater for production of biopesticides – Ramifications of solids concentrations. Environ Technol 30: 393-405.
19. Vu KD, Yan S, Tyagi RD, Valéro JR, Surampalli RY (2009c) Induced production of chitinase to enhance entomotoxicity of *Bacillus thuringiensis* employing starch industry wastewater as a substrate. Biores Technol 100: 5260-5269.
20. Vu KD, Adjallé KD, Tyagi RD, Valéro JR, Surampalli RY (2012a) *Bacillus thuringiensis* based-biopesticides production using starch industry wastewater fortified with different carbon/nitrogen sources as fermentation media. Unpublished results.
21. Vu KD, Adjallé KD, Tyagi RD, Valéro JR, Surampalli RY (2012b) Recovery of chitinases and Zwittermicin A from *Bacillus thuringiensis* fermented broth to produce biopesticide with high biopesticidal activity. Unpublished results.
22. Yezza A, Tyagi RD, Valéro JR, Surampalli RY (2005) Wastewater sludge pre-treatment for enhancing entomotoxicity produced by *Bacillus thuringiensis* var. *kurstaki*. World J Microbiol Biotechnol 21: 1165-1174.
23. Yezza A, Tyagi R.D., Valero JR, Surampalli RY (2006) Correlation between entomotoxicity potency and protease activity produced by *Bacillus thuringiensis var. kurstaki* grown in wastewater sludge. Process Biochem 41: 794-780.
24. Yezza A, Tyagi RD, Valéro JR, Surampalli R.Y (2006b) Bioconversion of industrial wastewater and wastewater sludge into Bacillus thuringiensis based biopesticides in pilot fermentor. Biores Technol 97: 1850-1857.
25. Zouari N, Jaoua S (1999a) Production and characterization of metalloproteases synthesized concomitantly with delta-endotoxin by

Bacillus thuringiensis subsp. *kurstaki* strain grown on gruel-based media. Enz Microb Technol 25: 364-371

26. Zouari N, Jaoua S (1999b) The effect of complex carbon and nitrogen, salt, Tween-80 and acetate on delta-endotoxin production by a *Bacillus thuringiensis* subsp *kurstaki*. J. Ind Microbiol Biotechnol 23: 497-502.

CHAPITRE 5. CONCLUSIONS ET RECOMMANDATIONS

5.1. Conclusions

Les eaux usées d'industrie d'amidon (SIW) produites dans le monde peuvent causer un problème de pollution de l'environnement si elles ne sont pas gérées efficacement. À ce sujet, au lieu de les traiter par les méthodes conventionnelles en aérobiose ou anaérobiose, au cours de nos travaux R&D nous avons recyclé biologiquement des SIW en produits à haute valeur ajoutée soit en biopesticides à base de *Bacillus thuringiensis*. Pour la formulation finale de ce produit à partir de suspensions du culot (obtenues à partir de surnageants du bouillon fermenté par centrifugation), une haute entomotoxicité (Tx) de la suspension du culot est nécessaire. Par conséquent, ce projet de recherche visait à développer des stratégies et des méthodes de production pour augmenter la Tx de la suspension du culot. En outre, les relations entre les différents paramètres (spores, delta-endotoxines et Tx) de biopesticides produits à partir de différents types de milieux ont été étudiées:

1. Tout d'abord, quatre méthodes / stratégies ont été appliquées pour accroître les éléments nutritifs des SIW accessibles pour l'obtention de Btk HD-1 pour atteindre une meilleure valeur de Tx contre la tordeuse des bourgeons de l'épinette (TBE): (1) enrichissement en nutriments (acide acétique, hydroxyde d'ammonium) dans des SIW au cours de la fermentation, sous la forme de différents agents de contrôle du pH; (2) utilisation de différentes concentrations en solides totaux de SIW pour la fermentation; (3) Addition de source(s) de carbone et d'azote dans SIW; et (4) fermentation en fed-batch. Une concentration significativement plus grande en delta-endotoxines, des valeurs plus élevées de Tx de quatre bouillons fermentés et de quatre suspensions des culots ont été obtenues par rapport aux résultats observés dans les contrôles (SIW sans modification). Les résultats obtenus et les conclusions tirées par l'application de ces méthodes sont les suivantes:

- Les SIW (sans modification) ne contiennent pas suffisamment de nutriments pour offrir le maximum potentiel insecticides (Tx) des biopesticides à base de Btk HD-1 dans le bouillon fermenté et dans la suspension du culot (Tx = 20.9 x 10^6 SBU/ml).
- Cependant, en ajoutant des sources de carbone, tels que le CH_3COOH (comme agent de contrôle du pH au cours de la fermentation) ou d'amidon de maïs mélangé avec du Tween 80 dans les SIW, le potentiel insecticide des suspensions du culot a augmenté (Tx = 26,7 x 10^6 SBU/ml) comparativement à les SIW initial (sans ajout de source de carbone).
- L'augmentation de la concentration des solides totaux de 15 g/L (présent dans les SIW recueillies à partir de l'industrie de la fabrication de l'amidon) de 30 g / L (par décantation) est une autre option pour augmenter les éléments nutritifs dans le milieu de fermentation pour ainsi augmenter la valeur de Tx de la suspension du culot (25,8 x 10^6 SBU/ml).
- La fermentation en Fed batch avec deux alimentations intermittentes à 10h et à 20h pourrait augmenter de façon significative les concentrations en cellules, en spores, en delta-endotoxines, de même que le potentiel insecticide de suspension du culot (27,4 x 10^6 SBU/ml).
- Il faut noter que l'augmentation de Tx du bouillon fermenté n'est pas proportionnelle à celle des concentrations en spores et en delta-endotoxines.

2. Deuxièmement, l'action synergique de chitinases vis-à-vis les delta-endotoxines et les spores afin d'augmenter la valeur de Tx a été étudiée. Les conclusions suivantes ont été tirées à partir des présents travaux de recherche:

- L'enrichissement des SIW avec de la chitine colloïdale (SIWC) augmente la Tx. La concentration optimale de la chitine colloïdale nécessaire pour obtenir les valeurs maximales de Tx du bouillon fermenté et de Tx de suspension du culot a été de 0,2% (p/v).

- En cas de fortification par de la chitine colloïdale, des valeurs maximales de Tx de bouillons fermentés (17.1 x 10^6 SBU/ml) et de Tx de suspensions du culot (26.7 x 10^6 SBU/ml) étaient significativement plus élevées que le maximum des valeurs de Tx du bouillon fermenté (15,3 x 10^6 SBU/ml) et de Tx de suspension du culot (20,5 x 10^6 SBU/ml) dans le contrôle (SIW sans chitine colloïdale); La valeur maximale de la Tx étant atteinte en 36 h de fermentation, au lieu de 48 h dans le contrôle (sans la chitine colloïdale).

3. Il a été constaté que le surnageant du bouillon de fermentation des SIWC contient de la Zwittermicine A (un antibiotique de 396 Da), ceci déterminé par des tests qualitatifs. Ainsi, les rejets des chitinases (et d'autres composants solubles tels que les protéines insecticides végétatives- Vips, spores résiduelle, delta-endotoxines résiduelle, etc...) et Zwittermicine A présents dans le surnageant du bouillon fermenté des SIWC (après centrifugation) pourrait entraîner une perte très importante de Tx. Afin de réduire au minimum cette perte et récupérer les composants solubles (pour augmenter la Tx), le surnageant a été soumis à l'ultrafiltration. Le processus d'ultrafiltration a fractionné les chitinases (et d'autres composants de haut poids moléculaire) dans le rétentat tandis que la Zwittermicine A (un antibiotique de 396 Da) a été retrouvée dans le filtrat, en utilisant une membrane ayant un seuil de coupure de 5 kDa. Le rétentat et le filtrat, ainsi obtenus, ont été mélangés à des suspensions du culot des SIWC à différents ratios. Les Tx des différents mélanges ont été évaluées par des bioessais contre la TBE et les conclusions furent les suivantes :

- Le mélange de suspension du culot avec rétentat dans un ratio volumétrique de 1:4 (1 L suspension du culot et 4 L de rétentat) a donné la plus haute récupération de la Tx (65,1%) du bouillon fermenté avec le plus gros volume de 5 L. Les concentrations en spores, en delta-endotoxines et l'activité des chitinases dans ce mélange étaient de 3,24 x 10^8 (CFU/ml), 1,09 (mg/ml) et 69,3 (mU/ml), respectivement. La valeur de Tx de ce mélange a été 22,8 x 10^6 (SBU/ml), ce qui était le ratio de mélange volumétrique le plus efficace en termes de synergie d'action des chitinases, des delta-endotoxines et des spores. Par conséquent, ce mélange peut être utilisé pour formuler les produits finaux.

- De même, le mélange de suspension du culot avec le filtrat dans un ratio volumétrique de 1:2 (1 L suspension du culot et 2 L filtrat) est le ratio de mélange volumétrique le plus efficace en termes de synergie d'action de la Zwittermicine (dans filtrat) avec les delta-endotoxines et les spores sur la Tx. Les concentrations en spores, en delta-endotoxines, l'activité des chitinases et la valeur de Tx de ce mélange atteignaient 5,40 x 10^8 (CFU/ml), 1,82 (mg/ml), 10,2 (mU/ml), et 22,8 x 10^6 (SBU/ml), respectivement. Cependant, ce mélange a été moins efficace en termes de récupération de la Tx du bouillon fermenté par rapport au mélange de 1 L de suspension du culot et 4 L de rétentat.

4. Enfin, six différents milieux basés sur des milieux semi-synthétiques (SYN); des eaux usées d'industrie d'amidon (SIW) et des boues d'épuration (SLUDGE) ont été utilisés pour l'obtention de Btk HD-1. Les données des concentrations en spores, en delta-endotoxines et des valeurs de Tx obtenus dans ces expériences ont été enregistrées et les relations entre ces paramètres ont été établies. Les conclusions suivantes ont été tirées:

- La concentration en delta-endotoxines augmente avec celle en spores. La relation entre ces paramètres suit une loi de puissance, indépendamment du milieu de fermentation utilisé pour cultiver Btk HD-1.
- La Tx spécifique à l'égard des spores (Tx par 1000 spores) a diminué avec l'augmentation de la concentration en spores et la relation entre la Tx et la concentration en spores suit aussi strictement une loi de

puissance, indépendamment du milieu utilisé; La caractéristique des constantes de la loi de puissance variant avec le type de milieu utilisé.

- La relation entre la Tx et la concentration en Delta-endotoxines a été jugée exponentielle, indépendamment des types du milieu utilisé. Par conséquent, il est possible de prévoir Tx en se basant sur la concentration en delta-endotoxines dans chaque milieu spécifique. Cependant, pour avoir la valeur exacte de chaque Tx des bouillons fermentés, des bioessais contre les insectes cible doivent être faits.

5.2. Recommendations

Voici quelques recommandations importantes pour les futures recherches (Tableau 23):

Tableau 23. Recommandations pour les futures recherches

Objectif	Méthodologie	Résultats attendus
Recherches sur les chitinases bactériennes comme agents de synergie		
1. Déterminer si la ou les chitinase(s) seules peuvent tuer la tordeuse des bourgeons de l'épinette, l'arpenteuse du chou ou autres insectes ravageurs	(a) production de chitinases par Btk HD-1 en utilisant des SIWC; (b) fractionnement et purification de chitinases en utilisant la chromatographie sur colonne; (c) estimation des poids moléculaires de chitinases par PAGE (électrophorèse en gel de polyacrylamide); (d) détermination de leurs activités (exo-chitinase, chitinase endo-, N-acetylglucosaminase) seules contre la tordeuse des bourgeons de l'épinette, arpenteuse du chou ou autres insectes nuisibles par des bioessais	L'activité insecticide de chitinases seules contre la tordeuse des bourgeons de l'épinette/arpenteuse du chou ou autres insectes nuisibles peut être déterminée. Si les chitinases seules n'ont pas la capacité de tuer la tordeuse des bourgeons de l'épinette, l'arpenteuse du chou et autres insectes nuisibles ou insectes utiles, puis la production de chitinases à grande quantité devrait être estimée

2. Production industrielle économique de chitinases en utilisant des eaux usées de crevettes/crabe comme matières premières	(a) Isolement de bactéries ou sélection de bactéries connaître qui peuvent produire des chitinases de hautes activités. b) Optimisation des processus pour la production de chitinases par l'utilisation des eaux usées de crevettes/crabe comme substrat de fermentation c) Un processus d'ultrafiltration pour le recouvrement de chitinases peut être employé.	De grandes quantités de chitinases pourraient être produites pour l'étape de pré-formulation .
3. Pré-formulation des biopesticides	Mélange de chitinases avec les suspensions du culot (obtenus à partir de la fermentation de Btk HD-1 en utilisant des eaux usées d'industrie d'amidon comme matières premières) à différents ratios pour savoir quel mélange a la plus haute efficacité contre certains insectes ravageurs	Un mélange (chitinases et suspension du culot) à haute efficacité contre les insectes ravageurs cible pourrait être obtenu.

D'autres recherches sur Zwittermicine A		
1. Production de Zwittermicine A à partir de *Bacillus cereus* UW85 utilisant des SIW ou des boues d'épuration comme milieux de fermentation	(a) Travaux préliminaires de production de Zwittermicine A par *Bacillus cereus* UW85 sur des SIW ou des boues. Au cours de la fermentation, des échantillons seront retirés et centrifugés. Le surnageant sera utilisé pour déterminer quantitativement la concentration de Zwittermicine A (b) Si *Bacillus cereus* UW85 peut produire Zwittermicine A sur ces milieux, de nouvelles recherches sur l'optimisation des processus de fermentation pour maximiser la production de Zwittermicine A peuvent être effectuées.	Paramètres optimisés pour la production de Zwittermicine A utilisant des SIW ou des boues comme substrats de fermentation
2. Récupération de Zwittermicine A	(a) Récupération de Zwittermicine A à partir de surnageant du bouillon fermenté par ultrafiltration en utilisant 1 kDa cut-off	De grandes quantités de Zwittermicine A pourraient être obtenues.

	membrane. La zwittermicine A sera présente dans le filtrat. (b) Lyophilisation du filtrat afin de produire le produit de Zwittermicine A. Détermination de la concentration de Zwittermicine A dans le produit final.	
3. Pré-formulation	(a) Addition de Zwittermicine A aux suspensions de culot (complexe spores et delta-endotoxines) à différents ratios pour déterminer le ratio du mélange permettant le maximum de Tx contre les insectes ravageurs cibles.	Un mélange (Zwittermicine A et suspension du culot) à haute efficacité contre les insectes ravageurs cibles pourrait être obtenu.
D'autres recherches sur la pré-formulation des chitinases et de la Zwittermicine A avec un complexe spores et delta-endotoxines		
1. Pré-formulation	L'ajout de chitinases avec la Zwittermicine A et suspension du culot à différents ratios pour identifier la action synergie des deux (chitinases et Zwittermicin A) avec spores et delta-	Un mélange (Zwittermicine A + chitinases + suspension du culot) à haute efficacité contre les insectes ravageurs cibles pourrait être obtenu.

	endotoxines sur insectes ravageurs cibles pour savoir quelle combinaison pourrait être la plus efficace en terme de Tx.	
Bioessais des (biopesticides finaux contre la fausse-arpenteuse du chou		
1. Détermination de la Tx de produits formulés contre la fausse-arpenteuse du chou.	Bioessais de produits formulés contre la fausse-arpenteuse du chou doivent être faits pour évaluer la puissance entomotoxique avant de les enregistrer comme produits commerciaux.	Homologation des produits finaux comme produits commerciaux.

MIX
Papier aus verantwortungsvollen Quellen
Paper from responsible sources
FSC® C105338

Printed by Books on Demand GmbH, Norderstedt / Germany